Remote Sensing of Ice and Snow

Remote Sensing of Ice and Snow

DOROTHY K. HALL
JAROSLAV MARTINEC

London New York
CHAPMAN AND HALL

First published in 1985 by
Chapman and Hall Ltd
11 New Fetter Lane, London EC4P 4EE
Published in the USA by
Chapman and Hall
29 West 35th Street, New York, NY 10001

Printed in Great Britain at the University Press, Cambridge

ISBN 0 412 25910 9

British Library Cataloguing in Publication Data

Hall, Dorothy K.
Remote sensing of ice and snow. —— (Remote sensing applications)
1. Glaciology 2. Remote sensing
I. Title II. Martinec, Jaroslav III. Series
551.3′1′028 QE576

ISBN 0-412-25910-9

Library of Congress Cataloguing in Publication Data

Hall Dorothy K., 1952–
Remote sensing of ice and snow.

Includes bibliographies and index.
1. Ice—remote sensing. 2. Snow—Remote sensing.
3. Frozen ground—Remote sensing. I. Martinec, J.
II. Title.
GB2401.72.R42H35 1985 551.3′1 85–9712
ISBN 0-412-25910-9

Contents

Colour plates appear between pages 118 and 119

Preface

Remote sensing using aircraft and satellites has helped to open up to intensified scientific scrutiny the cold and remote regions in which snow and ice are prevalent. In this book, the utility of remote sensing for identifying, mapping and analyzing surface and subsurface properties of worldwide ice and snow features is described. Emphasis is placed on the use of remote sensing for developing an improved understanding of the physical properties of ice and snow and understanding the interrelationships of cryospheric processes with atmospheric, hydrospheric and oceanic processes. Current and potential applications of remotely sensed data are also stressed.

At present, all-weather, day and night observations of the polar regions can be obtained from sensors operating in different portions of the electromagnetic spectrum. Because the approaches for analysis of remotely sensed data are not straightforward, Chapter 1 serves to introduce the reader to some of the optical, thermal and electrical properties of ice and snow as they pertain to remote sensing. In Chapter 2 we briefly describe many of the sensors and platforms that are referred to in the rest of the book. The remaining chapters deal with remote sensing of the seasonal snow cover, lake and river ice, permafrost, glacier ice and sea ice.

We would like to thank the individuals whose comments and reviews led to many improvements in the book: Dr Robert Bindschadler (NASA/Goddard Space Flight Center, Greenbelt, Maryland), Dr Jerry Brown (US Army Cold Regions Research and Engineering Laboratory, Hanover, New Hampshire), James Foster (NASA/Goddard Space Flight Center), Dr Robert Gurney (NASA/Goddard Space Flight Center), Dr Claire Parkinson (NASA/Goddard Space Flight Center), Donald Wiesnet (Satellite Hydrology, Inc., Vienna,

Virginia), Dr Richard Williams, Jr (US Geological Survey, Reston, Virginia) and Russell Wright (University of Maryland, College Park, Maryland). Thanks are also due to Joyce Tippett for typing the manuscript.

1

An introduction to the optical, thermal and electrical properties of ice and snow

1.1 Introduction

Remote sensing, defined as the measurement of properties of an object or feature on the Earth's surface by an instrument that is not in direct physical contact with the object or feature, enables scientists to obtain information about ice and snow in visible, near-infrared, thermal infrared, microwave, and other wavelengths (Fig. 1.1). Surface, near-surface and deep, subsurface regions of ice and snow features can be analyzed using remote sensing techniques. In this chapter, the optical, thermal and microwave properties are briefly reviewed as they pertain to the remote sensing of ice and snow. More detailed descriptions of optical, thermal and electrical properties of ice and snow can be found in Hobbs (1974), Glen and Paren (1975) and ASP (1983).

1.2 Optical and thermal properties of ice and snow

Spectral reflectivity of snow is dependent on snow parameters such as: grain size and shape, impurity content, near-surface liquid water content, depth and surface roughness as well as solar elevation (Choudhury and Chang, 1981 and NASA, 1982). Freshly fallen snow has a very high reflectance in the visible wavelengths as seen in Fig. 1.2. As it ages, the reflectivity of snow decreases in the visible and especially in the longer (near-infrared) wavelengths. This greater decrease in the near-infrared wavelengths is due largely to melting and refreezing within the surface layers and to the natural addition of impurities. Melting of snow increases the mean grain size and density by melting the smaller particles. Figure 1.3 illustrates the changes in snow reflectance with different snow-crystal radii as determined from model results of Choudhury and Chang

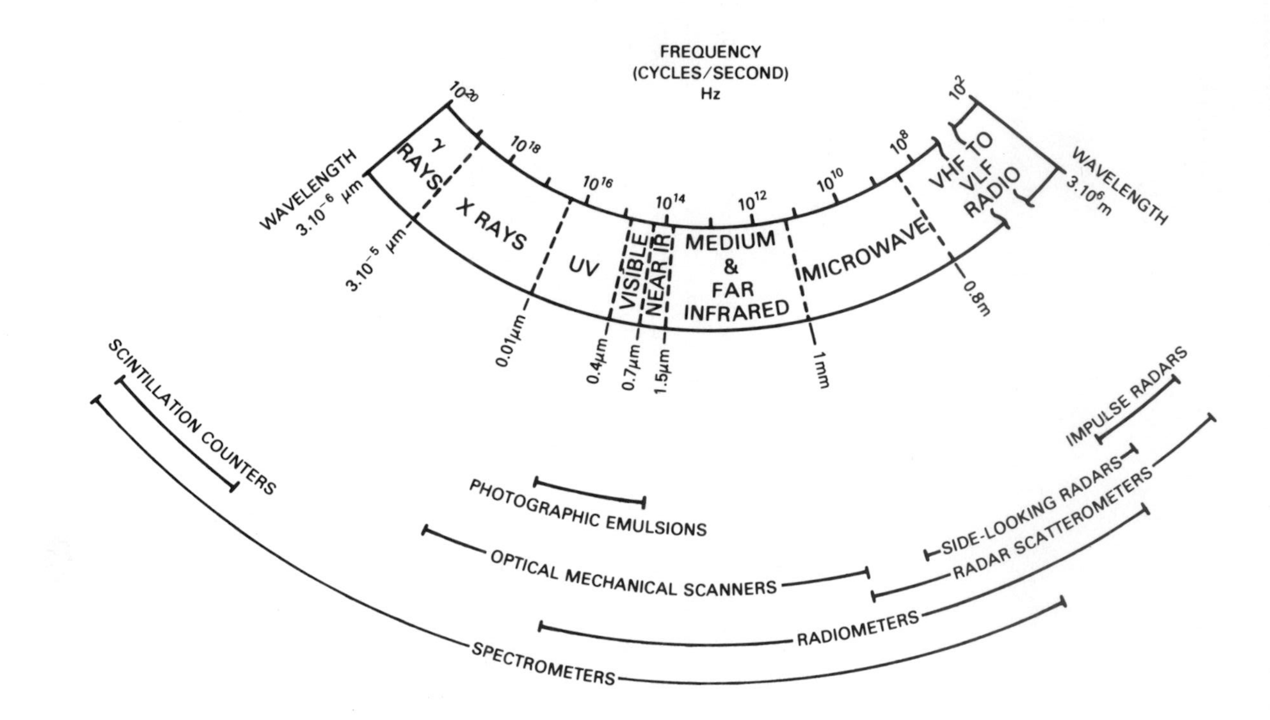

Fig. 1.1 Electromagnetic spectrum and instrumentation associated with specific electromagnetic intervals (adapted from Allison *et al.*, 1978).

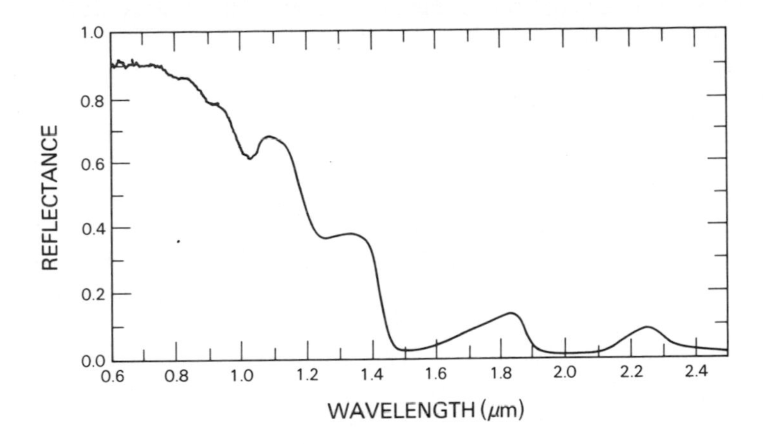

Fig. 1.2 Typical spectral reflectance curve for snow (adapted from O'Brien and Munis, 1975).

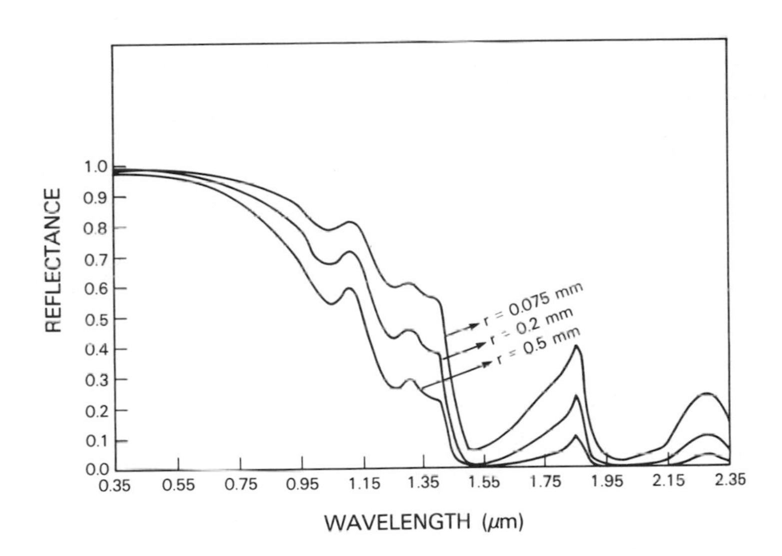

Fig. 1.3 Illustration of the effects of different snow crystal radii on snow reflectance as calculated by Choudhury and Chang (1979) (from Choudhury and Chang, 1979).

(1979). In the wavelength region between approximately 0.95 and 1.40 μm, differences in snow-crystal radius lead to the greatest differences in reflectivity as shown in Fig. 1.3.

There are several models that have been developed to calculate the reflectivity or albedo of snow. Formulation is complex and therefore not suitable for this chapter which is intended as an introduction only. The reader who would like further information on the snow reflectivity models is referred to the second edition of the *Manual of Remote Sensing*, Vol. 1, Chapter 3 in which three such models are compared (ASP, 1983).

Reflectivity of ice in the visible and near-infrared wavelengths from aircraft and satellites varies greatly, depending on the overlying material, impurities within the ice and the presence of surficial meltwater. For example, the reflectivity of glacier ice is quite low but glacier ice may be covered by snow and firn, especially in the accumulation zone, thus increasing the reflectivity. Meltwater in the lower zones and the presence of morainal material on the glacier surface tend to reduce the reflectivity. Thus, a wide range of reflectivities is known to characterize ice features as seen by remote sensors. Figure 1.4 exemplifies this variability of ice reflectance. Reflectance curves of snow, firn, glacier ice and dirty (moraine-covered) glacier ice are shown to vary greatly with wavelength. For example, fresh snow has a reflectance of almost 1.0 at 400 nm while its reflectance drops to 0.6 between 1000 and 1050 nm as seen in Fig. 1.4. The change in reflectance with wavelength is even more dramatic with glacier ice (as seen in Fig. 1.4), having a reflectance of approximately 0.65 at 600 nm, dropping to less than 0.1 at 1000 nm (Qunzhu *et al.*, 1984).

Thermal infrared sensors measure the radiation emitted by features in a wavelength region in which the physical temperature of an object can be calculated. The relationship of the wavelength of maximum emitted radiance, λ_{max}, and the absolute temperature, T, of the radiating object is given by Wien's Displacement Law which shows that the product of absolute temperature and wavelength is constant for a blackbody. (A blackbody is a perfect emitter at all wavelengths.

$$\lambda_{max} T = \omega \tag{1.1}$$

where ω is Wien's displacement constant. From Wien's Displacement Law one can see that the wavelength of maximum radiance increases with decreasing temperature. For the range of temperatures normally encountered on terrestrial surfaces, the wavelength range of maximum emitted radiance is between 8 and 14 μm with a correspondingly low reflectivity (Shafer, 1971).

Infrared radiant emission may be described by the Stefan–Boltzmann Law, which relates the total radiant emittance from a blackbody to the fourth power of the absolute temperature:

$$W = \epsilon \sigma T^4 \tag{1.2}$$

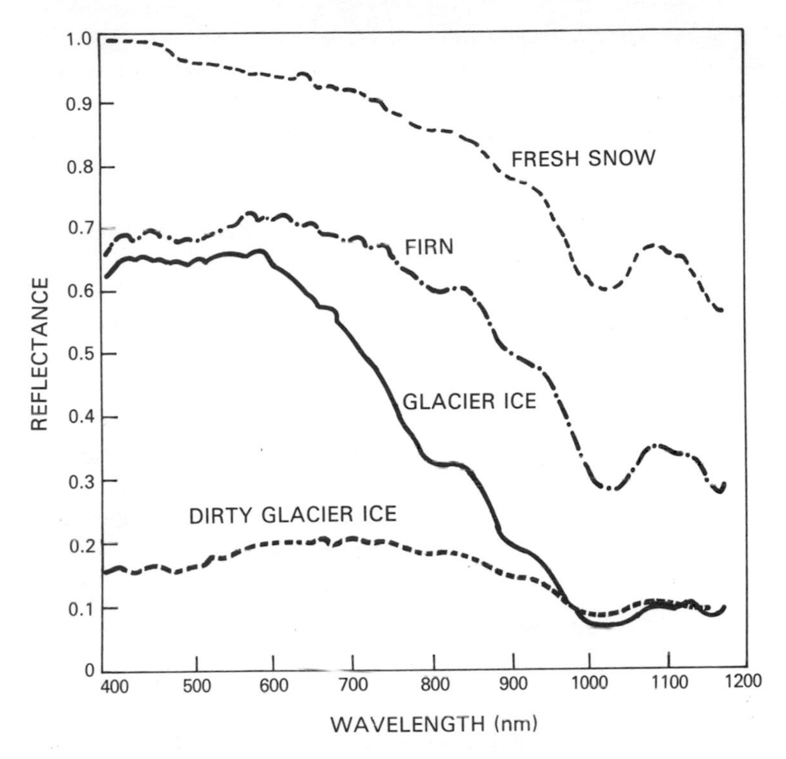

Fig. 1.4 Spectral reflectance curves for fresh snow, firn, glacier ice and dirty glacier ice. Note the extreme variability in the reflectance of ice and snow features (adapted from Qunzhu *et al.*, 1984).

where W is the radiant exitance (power per unit area), ϵ is the emissivity, σ is the Stefan–Boltzmann constant, and T is the absolute temperature (K). For non-blackbodies, the total radiant emittance from natural surfaces is less than 1.0 and ϵ_r would be substituted for ϵ in Equation (1.2) to represent the emissivity averaged over all wavelengths.

In addition to the absolute temperature of the feature, the quantity of radiant emission is affected by the emissivity as shown in Equation (1.2). Emissivity is a dimensionless number that represents the ratio of the radiant flux from a material to the radiant flux from a blackbody. Of the common natural surfaces found on the Earth, water is most like a blackbody (Short and Stuart, 1982). The emissivities of real materials on the Earth vary from 0 to less than 1.00 and are wavelength dependent.

In the case of snow, the temperature, crystal size and liquid water content affect the measurement of thermal infrared temperature. Snow depth can be important in a qualitative way because there must be a minimum thickness of snow in order not to allow radiation from below to affect the observed signature.

Another property that can be measured using thermal infrared remote sensing is the thermal inertia of a material. Thermal inertia, P, is the resistance of a material to temperature change:

$$P = (Kc\rho)^{1/2} \tag{1.3}$$

where K is thermal conductivity, c is specific heat and ρ is density (Short and Stuart, 1982).

The thermal infrared temperature that is measured by a remote sensor is modified by the atmosphere intervening between the sensor and the object. Thus, the atmospheric effects must be considered. The radiation measured when looking vertically downward is partly composed of reflected radiation from the sky in the clear sky situation. In the partially or wholly cloud-covered sky situation, the sky represents a much higher radiant temperature source than in the clear sky situation and, in partly cloudy skies, rapid change in the radiant emittance can occur (Shafer, 1971). These are important components of the thermal infrared radiant emittance of a remotely sensed feature.

1.3 Electrical properties of ice and snow

Microwave remote sensing can be accomplished either by measuring emitted radiation with a radiometer as seen in Fig. 1.5 or by measuring the intensity of the return (in decibels) of a microwave signal which has been sent, as with a radar. The electrical properties which govern the dielectric constant of a material, strongly influence the microwave emission and return. The complex dielectric constant is a measure of the response of a material to an applied electric

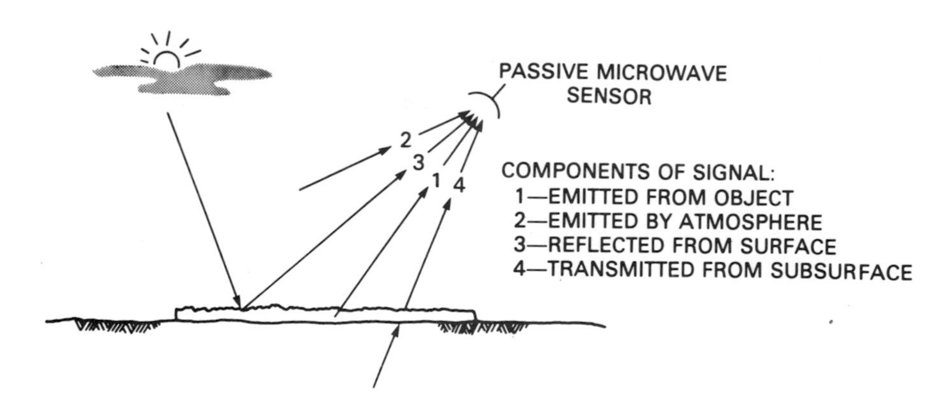

Fig. 1.5 Components of a passive microwave signal. Reprinted by permission of John Wiley & Sons, Inc. from *Remote Sensing and Image Interpretation*, T.M. Lillesand and R.W. Kiefer, Copyright © 1979, by John Wiley & Sons, Inc.

field such as an electromagnetic wave and is a function of temperature and frequency (Foster *et al.*, 1984). There are two components of the complex dielectric constant, ϵ, of a material: the real (ϵ') and the imaginary (ϵ''). Note that

$$\epsilon = \epsilon' - j\epsilon'' \quad (1.4)$$

where $j = (-1)^{1/2}$. The ratio ϵ''/ϵ' is called the loss tangent of the material.

Changes in the dielectric constant between layers within a medium cause reflection of an incident microwave signal or scattering of the natural microwave emission. To describe the electrical properties of snow, one must construct mixing formulas that consider ice and air, and water (if present). Presence of liquid water in the snow is the most important factor in changing the dielectric constant of snow.

In addition to dielectric changes, natural emission as detected by radiometers is affected by scattering in the bulk of the medium. Mie scattering, which has been used to describe the optical and microwave properties of clouds, has also been used to describe the microwave scattering in snow. For such models, the snow field is assumed to consist of randomly spaced scattering spheres (Fig. 1.6) which do not scatter coherently (Chang *et al.*, 1976).

Amount of penetration of microwaves through a medium is dependent upon the wavelength, crystal size, inclusions and impurities within the ice or snow

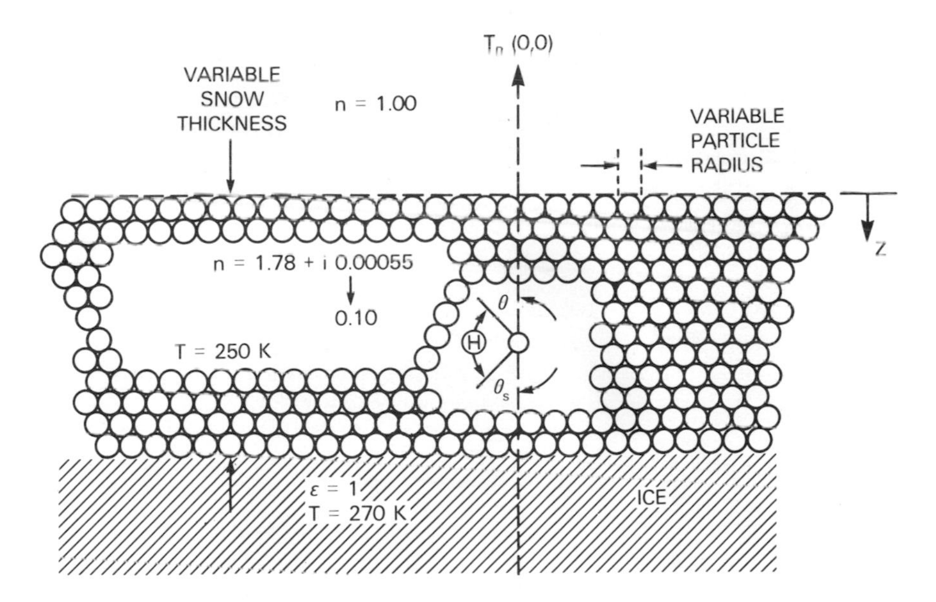

Fig. 1.6 Sketch of model snowfield overlying solid ice used for modelling of microwave emission from snow (adapted from Chang *et al.*, 1976).

medium. For radar, the depth of penetration of the incident radiation depends on its wavelength, τ, in the following way (ASP, 1983):

$$\delta = \left(\frac{\lambda}{\pi g \eta} \right)^{1/2} \tag{1.5}$$

where δ is the skin depth, g is the conductivity of the terrain and $\eta = (\mu/\epsilon)^{1/2}$, in which ϵ is the permittivity of the terrain and μ is the permeability of the terrain; the skin depth is defined as the depth below the surface at which the amplitude of the incident wave decreases to 37% of its value at the surface (ASP, 1983). Thus, in accordance with Equation (1.5), longer wavelengths allow deeper penetration.

In passive microwave remote sensing, Chang *et al.* (1976) show that the depth of penetration of upwelling microwave radiation through ice and snow is also wavelength dependent and that, depending upon the properties of the ice or snow, the emission can emanate from a depth that is 10 to 100 times the length of the wavelength. Thus, for a wavelength of 21.0 cm, depth of upwelling emission can range from 2.1 m to 21.0 m. As liquid water content in snow increases, the depth of penetration decreases.

For temperatures generally encountered on Earth, the emitted intensity of radiation is expressed as brightness temperature (T_B) in Kelvins and follows the Rayleigh–Jeans approximation which shows that the radiance from a blackbody is proportional to its temperature:

$$T_B = \epsilon T_s \exp - \tau + T_1 + (1 - \epsilon) T_2 \exp - \tau + (1 - \epsilon) T_{sp} \exp - 2\tau \tag{1.6}$$

where ϵ is the emissivity of the surface, T_s is the sensible temperature of the surface, τ is the total atmospheric opacity, T_1 is the upward emitted radiance contribution of the atmosphere, T_2 is the total downward (emitted and reflected) atmospheric brightness temperature, and T_{sp} is the averaged temperature of free space (Gloersen and Barath, 1977). For the purposes of this book, this equation can be simplified to:

$$T_B = \epsilon T_s \tag{1.7}$$

Many factors affect the radar returns and the brightness temperature measurements. These factors include instrument parameters such as wavelength and polarization as well as natural factors such as the presence of liquid water and surface roughness of the snow or ice feature.

References

Allison, L.J., Wexler, R., Laughlin, C.R. and Bandeen, W.R. (1978) Remote sensing of the atmosphere from environmental satellites. American Society for Testing and Materials, Philadelphia, PA, Special Publication 653, pp. 58–155.

ASP (American Society of Photogrammetry) (1983) *Manual of Remote Sensing*, 2nd edn (ed. R.N. Colwell), ASP, Falls Church, VA, Vols 1 and 2.

√ Chang, T.C., Gloersen, P., Schmugge, T., *et al.* (1976) Microwave emission from snow and glacier ice. *J. Glaciol.*, **16**, 23–39.

Choudhury, B.J. and Chang, A.T.C. (1979) Two-stream theory of reflectance of snow. *IEEE Trans. Geosci. Electron.*, **GE–17**, 63–8.

Choudhury, B.J. and Chang, A.T.C. (1981) The albedo of snow for partially cloudy skies. *Boundary-Layer Meteorol.*, **20**, 371–89.

Foster, J.L., Hall, D.K., Chang, A.T.C. and Rango, A. (1984) An overview of passive microwave snow research and results. *Rev. Geophys. Space Phys.*, **22**, 195–208.

Glen, J.W. and Paren, J.G. (1975) The electrical properties of snow and ice. *J. Glaciol.*, **15**, 15–38.

Gloersen, P. and Barath, F.T. (1977) A scanning multichannel microwave radiometer for Nimbus-G and Seasat-A. *IEEE J. Oceanic Eng.*, **OE–2**, 172–8.

Hobbs, P.V. (1974) *Ice Physics*, Clarendon Press, Oxford, England.

Lillesand, T.M. and Kiefer, R.W. (1979) *Remote Sensing and Image Interpretation*, John Wiley, New York.

NASA (1982) *Plan of Research for Snowpack Properties Remote Sensing – (PRS)^2. Recommendations of the Snowpack Properties Research Group*, NASA/Goddard Space Flight Center, Greenbelt, MD.

O'Brien, H.W. and Munis, R.H. (1975) Red and near-infrared spectral reflectance of snow. In *Operational Applications of Satellite Snowcover Observations* (ed. A. Rango), Proceedings of a workshop held in South Lake Tahoe, California, 18–20 August 1975, National Aeronautics and Space Administration, Washington, DC, NASA SP–391, pp. 345–60.

√ Qunzhu, Z., Meisheng, C., Xuezhi, F. *et al.* (1984) Study on spectral reflection characteristics of snow, ice and water of northwest China. *Sci. Sin.* (Series B), **27**, 647–56.

Shafer, B.A. (1971) *Infrared Temperature Sensing of Snow Covered Terrain*, MSc Thesis in Earth Science (Meteorology), Montana State University.

Short, N.M. and Stuart, L.M., Jr (1982) *The Heat Capacity Mapping Mission (HCMM) Anthology*, National Aeronautics and Space Administration, Washington, DC., NASA SP–465.

2

Sensors and platforms

2.1 Introduction

In this chapter, we will present a general description of most of the platforms and sensors that are referred to in this book. Sensors that operate in the gamma ray wavelengths to the very high frequencies (VHF) have been employed for remote sensing studies of ice and snow (Fig. 1.1). Although all objects emit radiation over all wavelengths, for most studies it is advantageous to use data from sensors operating in discrete portions of the electromagnetic spectrum. One must judiciously select the proper sensor to use for a particular analysis taking into consideration factors such as: wavelength, resolution and frequency and timing of ground coverage.

In general, satellite data from visible and near-infrared satellite sensors (e.g. from Landsat and NOAA satellites) are more readily available than other data, especially aircraft data. And the repetitive coverage of the Landsat and NOAA satellites is beneficial for many studies. But the sensors operating in the visible and near infrared wavelengths cannot produce an image through clouds. Thermal infrared data can be obtained at night but not through cloudcover. Microwave data (passive and active) can be obtained day or night and through most cloudcover but are presently more difficult to interpret than visible, near-infrared and thermal infrared data. This is because so many factors influence the microwave response including surface and subsurface factors, temperature, surface roughness and electrical properties of the features under study.

For more detailed information concerning the principles of remote sensing, the reader is referred to the second edition of the *Manual of Remote Sensing* (ASP, 1983) and the second edition of *Introduction to Environmental Remote Sensing* (Barrett and Curtis, 1982).

2.2 Multispectral Scanner (MSS) on the Landsat series

The Multispectral Scanner (MSS) has been flown continuously on five Landsat satellites since the launch of Landsat 1 in July 1972. (Note that Landsat 1 was called ERTS-1 until its name was changed in 1975.) The MSS is a line scanning device that uses an oscillating mirror to scan perpendicular to the orbital track of the spacecraft. Radiation is measured simultaneously by detectors in each of the following spectral bands:

Band 4 (0.5–0.6 μm)
Band 5 (0.6–0.7 μm)
Band 6 (0.7–0.8 μm)
Band 7 (0.8–1.1 μm) and
Band 8 (10.4–12.6 μm) – Landsat 3 only.

The output from the detectors is then sampled and formatted into a continuous data stream. The nominal instantaneous field of view (IFOV) for each detector in Bands 4, 5, 6 and 7 is approximately 80 m square.

MSS data are digitized on-board the spacecraft by an analog-to-digital converter and transmitted in a digital format to a ground receiving station when the satellite is within line-of-sight of a receiving station. When remote from a

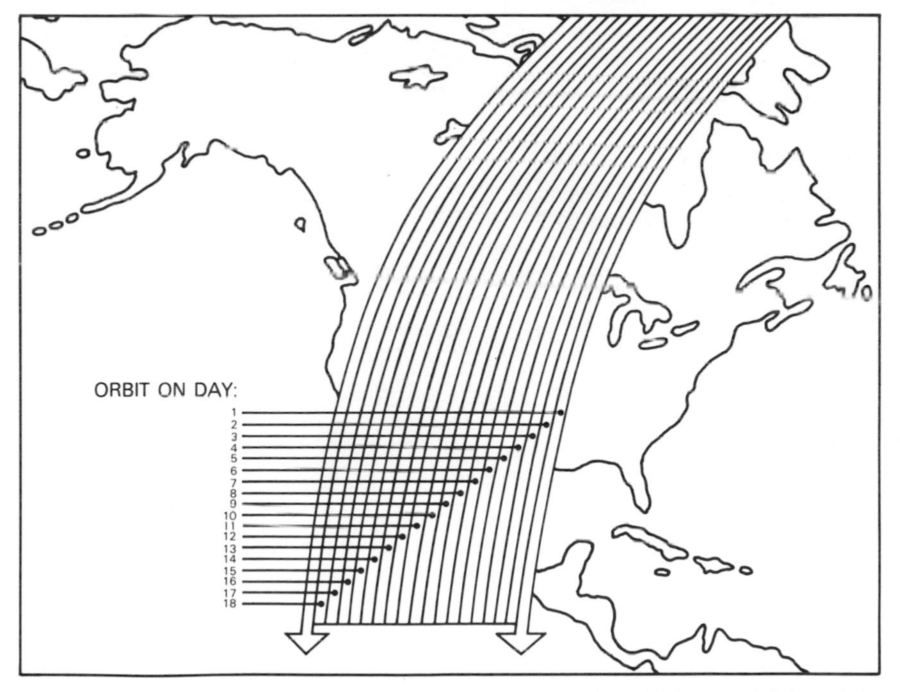

Fig. 2.1 Landsats 1, 2 and 3 swathing pattern (adapted from USGS and NOAA, 1984).

Table 2.1 Sidelap of adjacent Landsats 4 and 5 coverage swaths (USGS and NOAA, 1984)

Latitude (Degrees)	*Image sidelap (%)*
0	7.3
0	7.3
10	8.7
20	12.9
30	19.7
40	29.0
50	40.4
60	53.6
70	68.3
80	83.9

ground station, on-board tape recorders store data until the satellite passes over a ground station (Lillesand and Kiefer, 1979).

The Landsat satellites orbit the Earth progressing westward as can be seen in Fig. 2.1 which shows the Landsats 1, 2 and 3 swathing pattern. Continuous strip imagery is produced and transformed into framed images allowing for some overlap of portions of the ground between orbits. Overlap is greater at the higher latitudes than at the equator (Table 2.1). This is advantageous for ice and snow studies. Each frame represents a 185 km by 185 km area on the ground that is imaged once every 18 days for Landsats 1, 2 and 3 and once every 16 days for Landsats 4 and 5 (USGS, 1979). The Landsat orbit is sun-synchronous, and thus the satellite always crosses the equator at the same local (sun) time.

2.3 Thematic Mapper (TM) on Landsats 4 and 5

The Landsat 4 satellite was launched in July 1982 with an array of sensors having improved spatial resolution and radiometric sensitivity relative to the sensors on the previous Landsats. Landsat 5 has a sensor payload identical to that of Landsat 4. Thus, in addition to the MSS (as on Landsats 1, 2 and 3), Landsats 4 and 5 have a Thematic Mapper (TM) on-board with the following bands:*

Band 1 (0.45–0.52 μm)
Band 2 (0.52–0.60 μm)
Band 3 (0.63–0.69 μm)
Band 4 (0.76–0.90 μm)
Band 5 (1.55–1.75 μm)
Band 6 (10.40–12.50 μm) and
Band 7 (2.08–2.35 μm).

* Precise wavelength ranges may differ slightly.

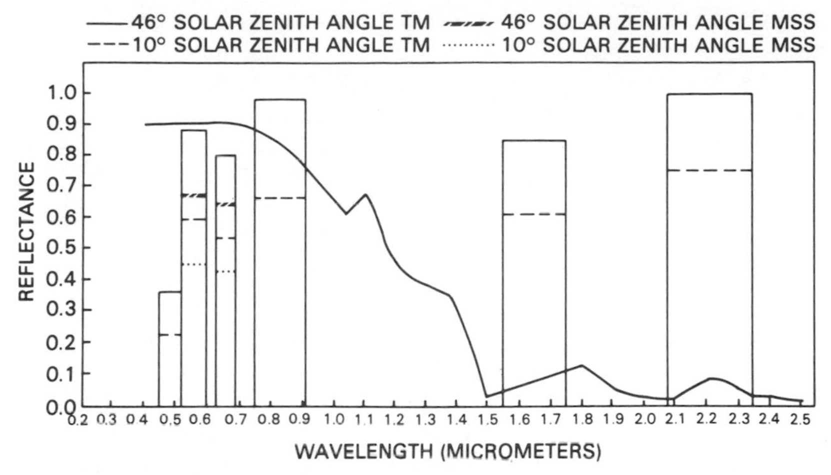

Fig. 2.2 Typical spectral reflectance curve for snow (O'Brien and Munis, 1975) showing saturation levels for visible and near-infrared bands of the thematic mapper (adapted from Salomonson and Hall, 1979).

Bands 1 through 5, and 7 are in the visible, near-infrared and middle infrared wavelength regions and have a spatial resolution of 30 m. Band 7, a thermal infrared band, has a resolution of 120 m. Rates for transmitting the data are considerably greater with Landsats 4 and 5 than with the previous Landsats. The Landsat 4 TM ceased operating in March 1984 and was replaced by the Landsat 5 TM.

Band 5 of the TM is especially useful for distinguishing between clouds and snow. The ability to distinguish between clouds and snow is desirable for accurate determination of the boundaries of snow in snow-cover mapping. Figure 2.2 shows the spectral reflectance of snow over the wavelength ranges of the TM. Note the dip in reflectance at approximately 1.5 μm. This permits the snow/cloud discrimination because the reflectance of clouds remains high.

2.4 NOAA satellites and sensors

NOAA (National Oceanographic and Atmospheric Administration) satellites and sensors that have been used for ice and snow studies and are discussed in this book are: NOAA 2, 3, 4, 5 and TIROS N. The NOAA satellites (including the TIROS series) operate in near-polar sun-synchronous orbits similar to the orbit of the Landsat series except at a higher altitude, allowing greater coverage per scene and more frequent coverage (twice daily), but poorer resolution. Daily visible and twice daily thermal infrared images are acquired by the NOAA satellites (Lillesand and Kiefer, 1979).

The VHRR (Very High Resolution Radiometer) was on-board the NOAA 2 through NOAA 5 satellites. The AVHRR (Advanced Very High Resolution Radiometer) was on-board TIROS-N and NOAA 6, and is currently operating on NOAA 7 and NOAA 8. The VHRR provided daily coverage in the visible (0.5–0.7 μm) region and twice daily coverage in the thermal infrared (10.5–12.5 μm) spectral region with a resolution of approximately 1 km at nadir. The AVHRR provides daily coverage in 4 or 5 channels (0.58–0.68 μm, 0.725–1.1 μm, 3.55–3.93 μm and 10.5–11.5 μm) at a resolution of 1.1 km. (On NOAA 7 and 8 the thermal infrared band is split into bands.)

The GOES (Geostationary Operational Environmental Satellite) series is placed in a fixed orbit over the equator at an altitude of 35 800 km. The satellite motion keeps the spacecraft fixed over a point on the equator. Distortion in GOES imagery increases with distance from the equator. Thus GOES data are of limited utility for high latitude ice and snow studies. The GOES VISSR (Visible and Infrared Spin–Scan Radiometer) produces visible (1 km resolution) and thermal (8 km resolution) images twice hourly. The visible VISSR sensor operates in that portion of the spectrum between 0.55 and 0.75 μm and the thermal sensor operates between 10.5 and 12.5 μm (McGinnis and Schneider, 1978).

2.5 Heat Capacity Mapping Mission (HCMM)

The Heat Capacity Mapping Mission (HCMM) satellite was operational from April 1978 until September 1980. The Heat Capacity Mapping Radiometer (HCMR) on-board HCMM was designed to obtain spectral information in the visible and near-infrared (0.55–1.1 μm) and thermal infrared (10.5–12.5 μm) wavelength regions. The imagery was produced with a nominal swath width of 720 km. The ground resolution beneath the spacecraft was 500 m on a side for the visible and near-infrared channels and 600 m for the thermal infrared channel. The HCMM orbit was sun-synchronous and covered areas between 85° north and 85° south latitude. Both day and night passes over selected areas at 12 hour intervals were repeated every 16 days (NASA, 1980). Because data were not recorded on-board the spacecraft, coverage of the Earth was limited to parts of the United States, Europe and Australia within range of one of five receiving stations: Merritt Island, Florida; Greenbelt, Maryland; Goldstone, California; Madrid, Spain; Orroral, Australia (Lillesand and Kiefer, 1979). Additional coverage, beginning in 1979, was acquired from stations in France and Alaska.

The HCMM was designed to investigate the feasibility of using the thermal inertia characteristics (see Chapter 1) of different materials to discriminate between different surface materials and to identify changing moisture conditions, e.g. soil moisture. Since the thermal inertia effect is related to the diurnal temperature variation, which has a maximum and minimum, the mission was designed to acquire thermal data on a temporal basis consistent with both maximum and minimum surface temperature variation.

2.6 Nimbus 5 and 6 Electrically Scanning Microwave Radiometer (ESMR) and Nimbus 7 Scanning Multichannel Microwave Radiometer (SMMR)

Passive microwave sensors have been flown on satellites for geophysical studies since 1972 when the Nimbus 5 satellite was launched. Before that and at present, passive microwave sensors have been and continue to be flown on aircraft for ice and snow studies. The sensors measure microwave radiation emitted by features on the Earth and the data generally are not adversely affected by non-precipitating clouds.

The Nimbus 5 satellite was launched in December 1972 into a sun-synchronous, circular orbit at an altitude of approximately 1100 km. The ESMR (Electrically Scanning Microwave Radiometer) (Wilheit, 1972), one of several instruments on-board the spacecraft, operated at a frequency of 19.35 GHz (1.55 cm) sensing horizontally polarized radiation at a resolution of approximately 25 km. The Nimbus 5 ESMR data record extends over a period of 11 years. High quality data were obtained for much of the 4 year period from December 1972 through December 1976 and lesser quality data were obtained after December 1976 through March 1983.

The Nimbus 6 ESMR (Wilheit, 1975) was launched in June 1975 and operated at a frequency of 37 GHz (0.81 cm), recording both horizontally and vertically polarized data at a resolution of approximately 30 km. The shorter wavelength of the Nimbus 6 ESMR (0.81 cm) as compared to the Nimbus 5 ESMR (1.55 cm) enabled a multispectral approach to be employed for the study of many ice and snow features (Table 2.2). ESMR data have been used

Table 2.2 ESMR instrument characteristics (Nimbus 5 and 6) (after ASP, 1983)

Characteristics	*ESMR (Nimbus 5)*	*ESMR (Nimbus 6)*
Frequency (GHz)	19.35	37
Wavelength (cm)	1.55	0.81
RF bandwidth (MHz)	250	250
Integration time (msec)	47	
Sensitivity, ΔT_{rms}(K)	1.5	
Dynamic range (K)	50–330	
Absolute accuracy (K) (longterm)	2	
IF frequency range (MHz)	5–125	
Antenna beamwidth (degrees)	1.4 × 1.4 (nadir position) 1.4 × 2.2 (scan extreme)	1.0 × 0.95 (scan center) 0.85 × 1.17 (scan extreme)

extensively to study the distribution and characteristics of snow cover over large areas (Chapter 4), the Greenland and Antarctic ice sheets (Chapter 7) and sea ice (Chapter 8).

In 1978 the Nimbus 7 satellite was launched with a Scanning Multichannel Microwave Radiometer (SMMR) on-board. The Nimbus 7 SMMR obtains dual-polarized data at five different wavelengths: 0.81, 1.43, 1.66, 2.80 and 4.54 cm. Some characteristics of the SMMR are summarized in Table 2.3. Further information on the SMMR and its characteristics can be obtained from Gloersen and Barath (1977). A wide range of snow and ice features can be analyzed using different frequencies of the SMMR. However, as with the Nimbus 5 and 6 ESMR data, studies of very large features such as large snowfields, ice sheets and sea ice are most compatible with the SMMR resolutions which range from 30 to 156 km.

Table 2.3 Some characteristics of the SMMR (after Gloersen and Barath, 1977)

Wavelength (cm)	0.81	1.43	1.66	2.80	4.54
Frequency (GHz)	37.00	21.00	18.00	10.69	6.60
Spatial resolution (km)	30	60	60	97.5	156
Temperature resolution t_{rms} (K) (per IFOV)	1.5	1.5	1.2	0.9	0.9
Antenna beam width (degrees)	0.8	1.4	1.6	2.6	4.2

2.7 Passive microwave aircraft sensors

On-board the Convair-990 aircraft (Fig. 2.3), which is a heavily instrumented NASA aircraft used for remote sensing, ESMR aircraft models and a SMMR simulator have been flown for geophysical studies of ice and snow. The resolution obtained by the aircraft models varies with altitude whereas the resolution from the satellite sensors is fixed because the orbital altitude is fixed. Thus, low-level flights offer increased spatial resolution but a narrower swath width on the resulting imagery as compared to higher altitude flights for which the resolution is poorer but the swath is wider.

The Multifrequency Microwave Radiometer (MFMR) is a non-imaging or non-scanning, four frequency radiometer which has been flown on the NASA P-3 and C-130 aircraft for ice and snow and other studies. The instrument obtains vertically and horizontally polarized data at the following frequencies: 37 GHz (0.81 cm), 22.2 GHz (1.4 cm), 18.0 GHz (1.7 cm) and 1.4 GHz (21.0 cm). The radiometers must be calibrated before each flight by a pass over water which has a known emissivity. Data are recorded as brightness temperatures (T_B) as with the ESMR and SMMR data.

Fig. 2.3 Photograph of the NASA Convair-990 aircraft in Fairbanks, Alaska (Photograph by D. Hall).

2.8 Synthetic Aperture Radar (SAR)

A radar image can be constructed by recording the intensity of microwave energy reflected or scattered from each resolution cell and received by the antenna. The intensity or brightness of an individual resolution cell is related to backscattered energy. More scattering will cause a brighter or stronger return. The surface parameters which have been found to influence the return signal are surface roughness, orientation, slope, and the complex dielectric constant (MacDonald and Waite, 1973). Incidence angle, polarization and frequency are the instrument parameters which also affect the intensity of the return signal. Figure 2.4 shows the geometry of a Side Looking Airborne Radar (SLAR).

Relative surface roughness may be calculated by using the smooth and rough criteria of Peake and Oliver (1971). These criteria are for smooth surfaces:

$$h < \frac{\lambda}{8 \sin \gamma} \tag{2.1}$$

for rough surfaces

$$h > \frac{\lambda}{4.4 \sin \gamma} \tag{2.2}$$

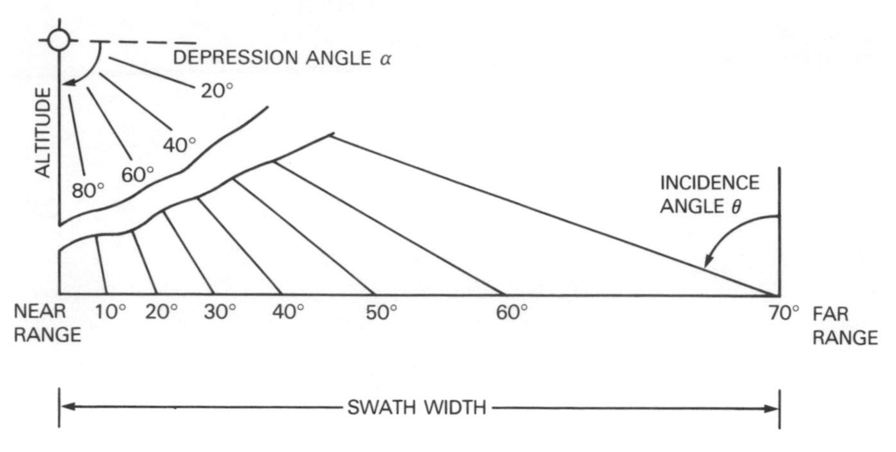

Fig. 2.4 Geometry of a Side Looking Airborne Radar (SLAR) (adapted from MacDonald and Waite, 1971).

where h is the average height of surface irregularities (in cm), λ is the radar wavelength, γ is the depression angle between the horizontal plane and the radar wave incident upon the terrain.

The length of the electromagnetic wave with respect to the size of the terrain feature determines whether a surface appears rough or smooth at a particular wavelength. A surface that is rough in the visible wavelengths may be quite smooth in the longer microwave wavelengths. A rough surface scatters the incident energy in all directions, returning some of it to the antenna. But a smooth surface or a specular reflector reflects the incident energy in one direction acting like a mirror. If the smooth surface happens to be perpendicular to the incident radar beam, then the energy returned to the antenna is intense. However, if the surface is at any other angle to the radar beam, none of the energy is received by the antenna.

Smooth water surfaces are excellent specular reflectors. Because they are normally not viewed at right angles to the SAR, they specularly reflect all the microwave energy into space. Thus, on SAR imagery rivers and lakes having smooth surfaces usually appear black. Conversely, related horizontal and vertical surfaces may work together to form a corner reflector, returning a large part of the energy directly back to the antenna. Such surfaces on SAR imagery appear much brighter than rough surfaces.

The wavelength of a radar signal influences the extent of penetration into the ground and the extent to which the signal is attenuated in the atmosphere (see Chapter 1). Radar wavelength ranges are often designated by bands with a letter as seen in Table 2.4.

Table 2.4 Radar band designations (after Lillesand and Kiefer, 1979)

Band	*Wavelength (cm)*	*Frequency (GHz)*
Ka	0.75 – 1.1	40.0 – 26.5
K	1.1 – 1.67	26.5 – 18.0
Ku	1.67 – 2.40	18.0 – 12.5
X	2.40 – 3.75	12.5 – 8.0
C	3.75 – 7.50	8.0 – 4.0
S	7.50 – 15.0	4.0 – 2.0
L	15.0 – 30.0	2.0 – 1.0
P	30.0 – 100.0	1.0 – 0.3

2.9 Seasat SAR and radar altimeter

The Seasat spacecraft was launched in June 1978 into a nearly circular, near-polar 800 km orbit. Seasat orbited the Earth 14 times a day at an inclination angle of 108° until the satellite experienced a massive short circuit in October of 1978 and was no longer operable. The synthetic aperture radar (SAR) and the radar altimeter were two of the five instruments on-board.

The Seasat SAR (Table 2.5), a synthetic aperture side looking imaging radar, transmitted a pulsed electromagnetic wave at a wavelength of 23.5 cm (1.28 GHz) in a horizontally like-polarized mode. Images can be constructed from the signals received from the target scene. The intensity of radar backscatter is affected by the surface and near-surface physical and electrical properties of the target scene (Elachi, 1980; Fu and Holt, 1982; Pravdo *et al.*, 1983).

Table 2.5 Some characteristics of the Seasat SAR (after Fu and Holt, 1982)

Parameter	*Value*
Satellite altitude	800 km
Radar frequency	1.275 GHz
Radar wavelength	23.5 cm
System bandwidth	19 MHz
Theoretical resolution on the surface	25 m × 25 m
Number of looks	4
Swath width	100 km
Antenna dimensions	10.74 m × 2.16 m
Antenna look angle	20° from vertical
Incidence angle on the surface	23° ± 3° across the swath
Polarization	HH

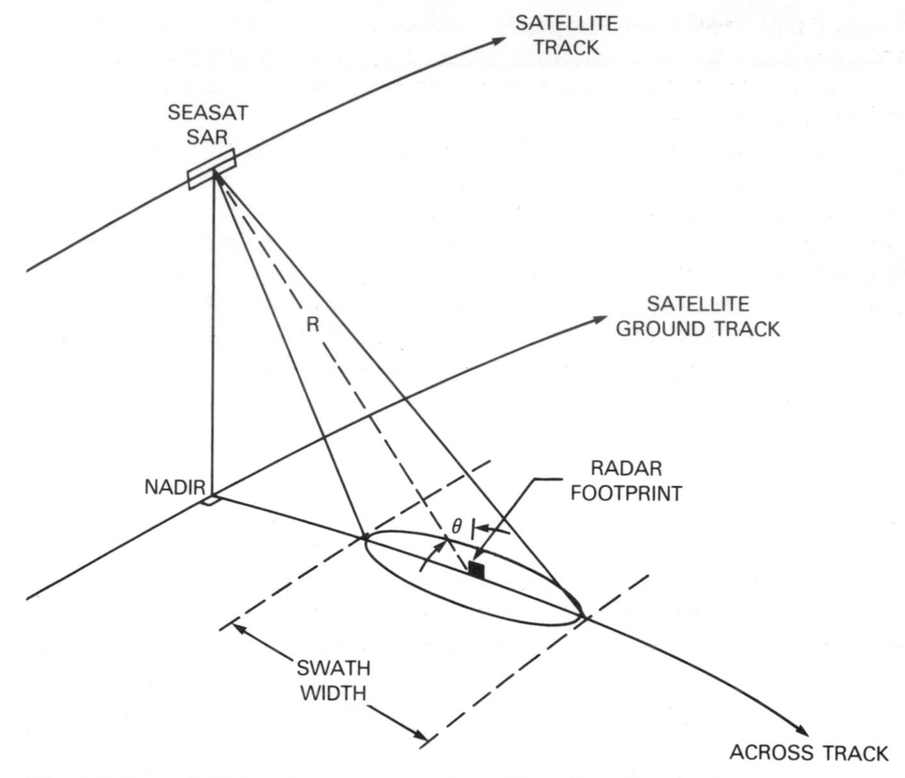

Fig. 2.5 Seasat SAR imaging geometry (adapted from Fu and Holt, 1982).

The Seasat SAR (Fig. 2.5) had a look angle of approximately 20° and incidence angles (measured normal to terrain surfaces) that commonly ranged between 0° and 30°. Surfaces having a slope greater than 20° and normal to the look direction produced geometrical distortion on the resulting imagery due to the layover effect (Wu *et al.*, 1981).

In Fig. 2.5, R is the distance between the radar and the target and is called slant range and θ is the incidence angle of the radar pulse. The image resolution in the across track direction, d_r, is determined by the effective pulse width, τ (Fu and Holt, 1982):

$$d_r = c\tau/2\sin\theta \tag{2.3}$$

where c is the speed of light.

Most of the Seasat SAR data were optically rather than digitally processed. In optical processing, a signal is recorded on film. Then the signal film data is transferred onto image film by passing a parallel beam of coherent light through a series of lenses onto the image film.

The Seasat SAR used the satellite motion to synthesize a large aperture antenna. The resolution obtainable is 25 m × 25 m and the imaged area is 100 km wide and 4000 km long. Images normally were produced to depict 100 km × 100 km areas on the ground.

Also on-board Seasat was a radar altimeter that generated a 13.56 GHz signal. It was designed to be used over the oceans to measure surface roughness. However, it has also been used to measure surface elevations of the ice sheets (see Chapter 7). Outputs of the Seasat radar altimeter were: radar return pulse strength, ocean surface significant wave height, radar return pulse shape or altimeter waveform, and spacecraft-indicated range to the Earth's surface (Martin *et al.*, 1983). The data were temporarily recorded on-board the spacecraft and later recorded on magnetic tape after being transmitted to telemetry stations.

2.10 Impulse radar

VHF (very high frequency) impulse radars have been useful for numerous ice studies: freshwater ice thickness determination (Chapter 5), determination of ice and structure in permafrost (Chapter 6), and measurement of thickness, bottom topography and internal structure of glaciers and ice sheets (Chapter 7). Impulse radars have been operated from the ground and on airplanes and helicopters. The radiating signal penetrates into the ground or ice medium to an extent that is dependent upon the wavelength and the properties of the material. Reflections occur when an electromagnetic wave encounters an interface representing a change in electrical properties of the materials. Figure 2.6 illustrates a received radar signal with an accompanying graphic recorder display. The graphic record displays signal level versus time. The intensity of the signal return is indicated by the intensity of the gray level on the graphic display. Energy returning to the antenna is detected and displayed versus delay time.

The radars used for impulse radar sounding operate at lower frequencies (longer wavelengths) than do conventional radars (Fig. 1.1). This enables greater depth of penetration than is possible at the higher frequencies.

In order to determine the velocity of the wave in the ice or soil, the dielectric constant, ϵ, of the medium must be known. A measure of the travel time, t, in nanoseconds, of the incident radar wave can then be related to thickness, X, of homogeneous materials, for example, lake ice (Cooper *et al.*, 1976):

$$X = 14.99t/(\epsilon)^{1/2} \tag{2.4}$$

For lake ice with a dielectric constant of 3.1, then:

$$X = 8.51t \tag{2.5}$$

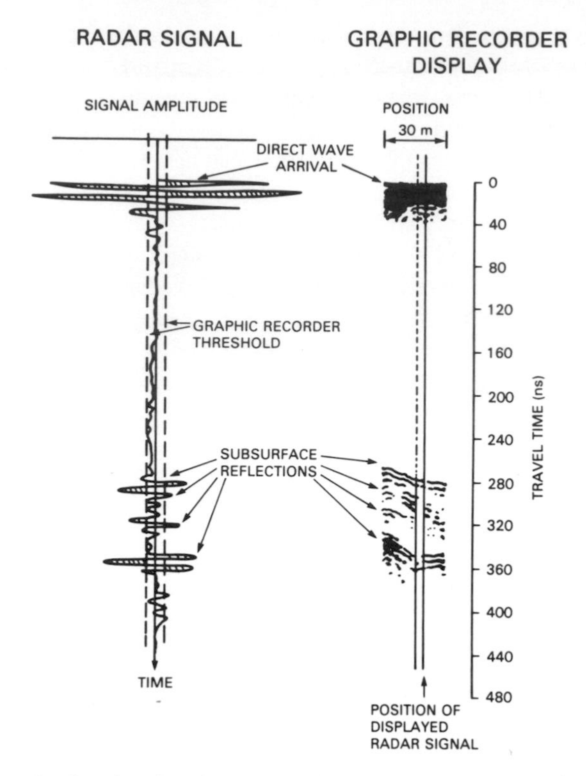

Fig. 2.6 A received radar signal accompanied by its graphic recorder display (from Annan and Davis, 1976).

References

Annan, A.P. and Davis, J.L. (1976) Impulse radar sounding in permafrost. *Radio Sci.* **11**, 383–94.

ASP (American Society of Photogrammetry) (1983) *Manual of Remote Sensing*, 2nd edn (ed. R.N. Colwell), ASP, Falls Church, VA, Vols 1 and 2.

Barrett, E.C. and Curtis, L.F. (1982) *Introduction to Environmental Remote Sensing*, 2nd edn, Chapman and Hall, London, England.

Cooper, D.W., Mueller, R.A. and Shertler, R.J. (1976) Remote profiling of lake ice using an S-band short-pulse radar aboard an all-terrain vehicle. *Radio Sci.* **11**, 375–81.

Elachi, C. (1980) Spaceborne imaging radar: geologic and oceanographic applications. *Science*, **209**, 1073–82.

Fu, L. and Holt, B. (1982) *Seasat Views Oceans and Sea Ice with Synthetic-Aperture Radar*, NASA Jet Propulsion Laboratory, JPL Publication 81–120.

Gloersen, P. and Barath, F. (1977) A Scanning Multichannel Microwave Radiometer for Nimbus-G and Seasat-A. *IEEE J. Oceanic Eng.*, **OE-2**, 172–8.

Lillesand, T.M. and Kiefer, R.W. (1979) *Remote Sensing and Image Interpretation*, John Wiley, New York.

MacDonald, H.C. and Waite, W.P. (1971) Optimum radar depression angles for geological analysis. *Mod. Geol.*, **2**, 179–93.

MacDonald, H.C. and Waite, W.P. (1973) Imaging radars provide terrain texture and roughness parameters in semi-arid environments. *Mod. Geol.*, **4**, 145–58.

Martin, T.V., Zwally, H.J., Brenner, A.C. and Bindschadler, R.A. (1983) Analysis and retracking of continental ice sheet radar altimeter waveforms. *J. Geophys. Res.*, **88**, 1608–16.

McGinnis, D.F., Jr, and Schneider, S.R. (1978) Monitoring river ice break-up from space. *Photogramm. Eng. and Remote Sensing*, **44**, 57–68.

NASA (National Aeronautics and Space Administration) (1980) *Heat Capacity Mapping Mission (HCMM) Data Users' Handbook for Applications Explorer Mission-A (AEM)*, NASA/Goddard Space Flight Center, Greenbelt, MD.

O'Brien, H.W. and Munis, R.H. (1975) Red and near-infrared spectral reflectance of snow. In *Operational Applications of Satellite Snowcover Observations* (ed. A. Rango), Proceedings of a workshop held in South Lake Tahoe, California, 18–20 August 1975, National Aeronautics and Space Administration, Washington, DC, NASA SP-391, pp. 345–60.

Peake, W.H. and Oliver, T.L. (1971) *The Response of Terrestrial Surfaces at Microwave Frequencies*, Ohio State University Electroscience Laboratory, Technical Report AFAL-TR-70-301.

Pravdo, S.H., Huneycutt, B, Holt, B.M. and Held, D.N. 1983: *Seasat Synthetic-Aperture Radar Data User's Manual*, NASA Jet Propulsion Laboratory, JPL Publication 82–90.

Salomonson, V.V. and Hall, D.K. (1979) A review of Landsat-D and other advanced systems relative to improving the utility of space data in water-resources management. In *Operational Applications of Satellite Snowcover Observations* (eds A. Rango and R. Peterson), Proceedings of a final workshop held at Sparks, Nevada, 16–17 April 1979, National Aeronautics and Space Administration, NASA CP-2116, pp. 281–96.

USGS (United States Geological Survey) (1979) *Landsat Data Users Handbook*, revised edn, US Department of the Interior, Sioux Falls, SD.

USGS and NOAA (1984) *Landsat 4 Data Users Handbook*, US Department of Commerce, Washington, DC.

Wilheit, T. (1972) The Electrically Scanning Microwave Radiometer (ESMR) experiment. In *Nimbus-5 User's Guide* (ed. R.R. Sabatini), pp. 59–103.

Wilheit, T. (1975) The Electrically Scanning Microwave Radiometer (ESMR) experiment. In *Nimbus-6 User's Guide* (ed. J.E. Sissala), pp. 87–108.

Wu, C., Barkan, B., Huneycutt, B. *et al.* (1981) *An Introduction to the Interim Digital SAR Processor and the Characteristics of the Associated Seasat SAR Imagery*, NASA Jet Propulsion Laboratory, JPL Publication 81–26.

3

Snow cover

3.1 Snow cover in the global water balance

According to estimates (Hoinkes, 1967), snow represents about 5% of all precipitation reaching the Earth's surface. This figure is rather uncertain because of difficulties in measuring solid precipitation by standard rain gages. By any standards, the volume of water stored as ice or snow on the Earth's surface is impressive. Table 3.1 gives estimates by Volker (1970) and by Baumgartner and Reichel (1975) of the world-wide distribution of fresh water.

It is of course the case that the total fresh water volume represents only 2.6% of all water in the hydrosphere. The remaining 97.4% is salt water in the oceans

Table 3.1 Distribution of fresh water on Earth

	Water volume			
	(10^6 km^3)		*(as percentage)*	
Forms of water presence	*a*	*b*	*a*	*b*
Polar ice, glaciers	24.8	27.9	76.93	77.24
Soil moisture	0.09	0.06	0.28	0.17
Groundwater within reach	3.6	3.56	11.17	9.85
Deep groundwater	3.6	4.46	11.17	12.35
Lakes and rivers	0.132	0.127	0.41	0.35
Atmosphere	0.014	0.014	0.04	0.04
Total	32.236	36.121	100	100

a: Volker (1970); b: Baumgartner and Reichel (1975)

and seas. Ice represents three times more fresh water than the total volume of fresh water in rivers, lakes and groundwater reservoirs. In addition to this seemingly permanent storage, water is each year temporarily stored in the seasonal snow cover which subsequently melts. Plate I shows areas with permanent, temporary and no snow cover. The seasonal snow cover is present in many industrialized countries where water is especially important for power generation, irrigation and municipal supply. Consequently, the seasonal snow cover is economically more important than glacier ice although the volume of water is considerably less. In terms of water supply, the average total flow of the world's rivers is about 36×10^3 km^3 in a year. Since evaporation losses are smaller from snow than from rain, it is possible that snow is responsible for about 10% of this volume. The total annual discharge of glaciers to the seas is estimated (Kotlyakov, 1970) at 2500–3000 km^3 of water. However, most of this volume is represented by iceberg calving. Thus it is not available as fresh water unless the idea of transporting icebergs to arid areas (Donaldson, 1978) becomes more realistic. Even then the transported water volume would be negligibly small in comparison with the river flow.

Nevertheless, the present total volume of water stored in ice is 700 times greater than the volume of the annual river runoff. About 99% of all freshwater ice is in Antarctica and Greenland (Mellor, 1964). Should this ice melt in a warmer climate resulting from the much discussed greenhouse effect of CO_2 in the atmosphere, the water level in the oceans would rise by about 70 m so that vast areas and some of the world's great cities would disappear under water. As for the rivers, ice on all continents except Antarctica, Greenland and Iceland could increase the present average flow by 10% for 50 years before gradually melting.

However, while equilibrium conditions between the renewal and melting are still more or less maintained, ice can be considered as a frozen deposit while seasonal snow represents an annual income. This difference is illustrated by the residence time:

$$t_r = \frac{V}{Q_e} \tag{3.1}$$

where t_r is the average residence time in years, V is the volume of water in the given medium in km^3 and Q_e is the rate of exchange by input (renewal by precipitation) and output (melting) in km^3 year^{-1}.

Assuming equilibrium between input and output, we roughly obtain for ice

$$t_r = \frac{25 \times 10^6\,\text{km}^3}{2500\,\text{km}^3\,\text{year}^{-1}} = 10\,000\,\text{years}$$

In contrast to this residence time, which is even longer for the Antarctic ice, snow stays a few weeks or months depending on the climate and elevation above sea level (a.s.l.). Figure 3.1 shows how the snowmelt influences the average

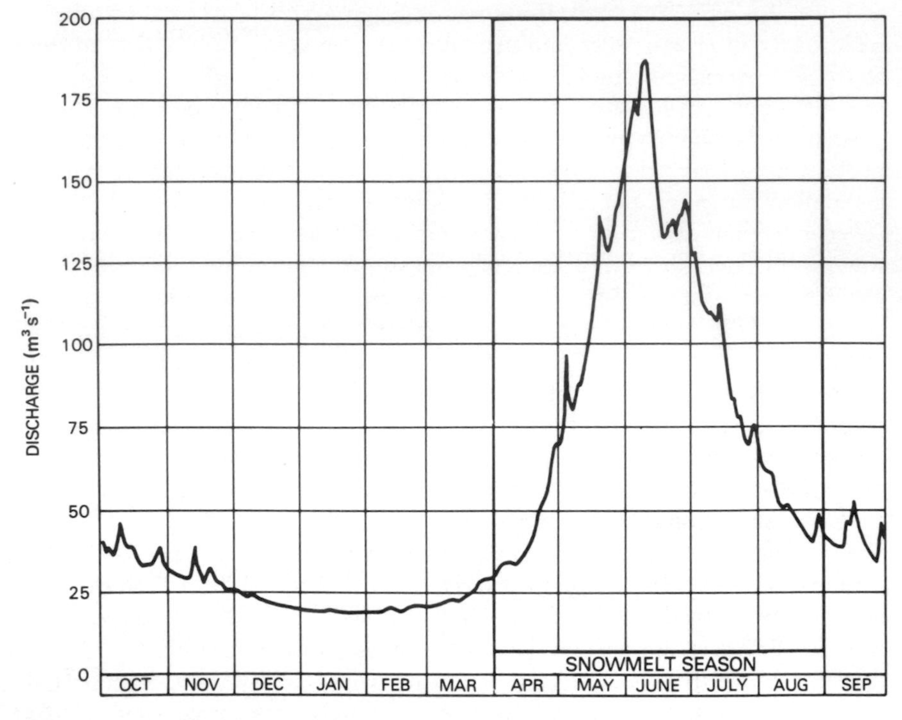

Fig. 3.1 Annual distribution of runoff in the Durance River basin, French Alps, 10-year averages for daily flows (hydrological years 1970–79).

annual runoff pattern in the Durance basin (French Alps, 2170 km^2, 786–4105 m a.s.l.). The presence of an important hydroelectric plant in this basin may serve as a hint of the economic importance of snow.

When the new technology of remote sensing became available for studying the Earth's surface, snow quickly attracted much attention because of its high albedo as compared with snow-free areas. However, it soon became evident that, in addition to the visible portion of the spectrum, the thermal infrared radiation as well as the microwave portion can provide more information about the snow cover. For this reason, it is necessary to take into account the properties of snow which are briefly outlined in the next section.

3.2 Snow properties

Snow is defined as falling or deposited ice particles formed mainly by sublimation (Unesco/IASH/WMO, 1970). Garstka (1964) describes snow as the solid form of water which grows while floating, rising or falling in the free air of the atmosphere. The simplest definition is from Bader (1962): snow is a porous, permeable aggregate of ice grains. According to a more recent definition (GHO, 1982), snow is solid precipitation composed of ice crystals, falling in the air or deposited on the ground.

Fresh snow ('new' snow in many other languages) is freshly deposited snow which has not been subjected to compaction or metamorphism. The depth of fresh snow is defined as referring to snow fallen in the last 24 hours. Consequently, fresh snow can already be partially compacted if there is a strong wind during snowfall. In new snow, the original shape of crystals is still recognizable. After deposition, metamorphism sets in and old snow consists of rounded or angular grains.

Firn is snow which has existed through at least one summer season and is carried over to the next winter. Alpine firn originates in conditions of repeated melting and refreezing while polar firn is created without appreciable melting.

The difference between all types of snow on the one hand and ice on the other hand is that snow has a connected system of air pores. Ice has air enclaves or closed air pores and a density exceeding 0.82 g cm^{-3}.

Typical values of some physical properties of snow and ice are given in Table 3.2 (Meier, 1964).

The density of snow is hydrologically a very important property since it enables the depth of snow to be converted into the water equivalent. It can be expressed as the mass per unit volume (g cm^{-3}, kg m^{-3}) or as a dimensionless relative density ρ_{rel}:

$$\rho_{rel} = \frac{\rho_s}{\rho_w} \tag{3.2}$$

where ρ_w is the density of water, 1 g cm^{-3} or 1000 kg m^{-3}.

The porosity mentioned in Table 3.2 is the volume ratio of voids to snow and is related to density as follows:

$$n = \frac{\rho_i - \rho}{\rho_i} = 1 - \frac{\rho}{0.917} = 1 - 1.09\rho \tag{3.3}$$

where n is the porosity expressed as a dimensionless ratio, ρ is the density of snow in g cm^{-3} and ρ_i is the density of ice in g cm^{-3}.

Snow depth is the vertical distance from the ground to the surface of the snow cover.

Water equivalent of snow is the vertical depth of a water layer which would be obtained by melting the snow cover.

The thickness of the snow cover should not be confused with the snow depth because it is measured perpendicularly to the slope.

As noted from Table 3.2, the density of snow increases with its age. This process can be accelerated by strong wind, warm temperature (Diamond and Lowry, 1953) and intermittent melting. However, time appears to be a dominant factor so that it is possible to derive a simple approximative relation (Martinec, 1977)

Table 3.2 Some physical properties of snow and ice

	Density ($g\,cm^{-3}$)	*Porosity* (%)	*Air permeability* ($g\,cm^{-2}\,s^{-1}$)	*Grain size* (*mm*)
New snow	0.01–0.3	99–67	>400–40	0.01–5
Old snow	0.2 –0.6	78–35	100–20	0.5 –3
Firn	0.4 –0.84	56– 8	40– 0	0.5 –5
Glacier ice	0.84–0.917	8– 0	0	1 –>100

$$\rho_n = \rho_0 (n + 1)^{0.3} \tag{3.4}$$

where ρ_0 is 0.1 g cm^{-3} which corresponds to the average density of new snow and ρ_n is the snow density after *n* days.

Thus, the density of new snow, $\rho_0 = 0.1$ g cm^{-3}, will increase after 1 day to $\rho_1 = 0.123$ g cm^{-3}, after 30 days to $\rho_{30} = 0.28$ g cm^{-3} and after 100 days to $\rho_{100} = 0.4$ g cm^{-3}.

Of course the initial density may deviate from the assumed value of $\rho_0 = 0.1$ g cm^{-3}. But the differences diminish rapidly with time as is shown in Fig. 3.2. If, for example, a snowfall is accompanied by a strong wind so that the

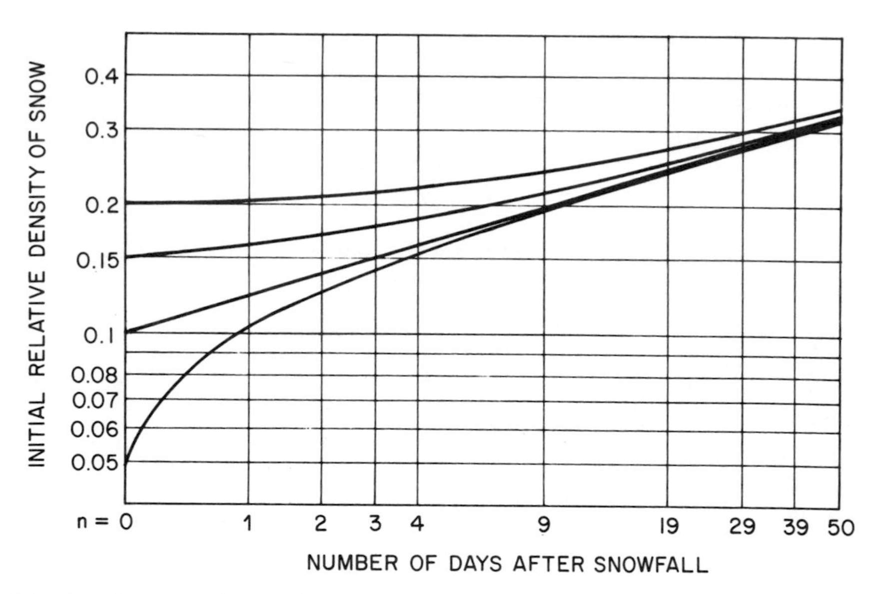

Fig. 3.2 Increase of snow density with time for snow layers with different initial densities.

new snow assumes a density of 0.2 g cm^{-3}, this relatively high value will become nearly normal after 50 days of average wind conditions. If a snowfall takes place on a day favoring a low density of snow (no wind, extremely low temperature), a density of $\rho_0 = 0.05$ g cm^{-3} will quickly approach values calculated by Equation (3.4) if the period following the snowfall corresponds to average conditions from which Equation (3.4) was derived.

Since direct measurements of the water equivalent of snow are sparse, this method is used to determine the water storage in a basin or snow loads on roofs (Martinec, 1977) if only snow depth data are available. It can also be used to obtain at least approximate ground truth data for remotely sensed water equivalents. Alternatively, if snow depths can be measured by remote sensing, it can convert these data into the desired water equivalents.

Albedo is a property of snow especially suitable to be remotely sensed. It is the ratio of the reflected to the incoming global radiation. It amounts to 90% or even more for a freshly fallen snow cover and drops below 40% if the snow surface is weathered and dirty. Values as low as 20% have been measured on an avalanche deposit. Consequently, fresh snow is easily recognized on the satellite imagery while old snow cover may at times look darker than certain snow-free surfaces. Figure 3.3 illustrates the decrease of albedo with time in comparison with the increase in snow density. Thus, a remotely sensed albedo could serve as an approximate index of the density of the top snow layer.

While it is not possible to deal with all the properties of snow and ice, the

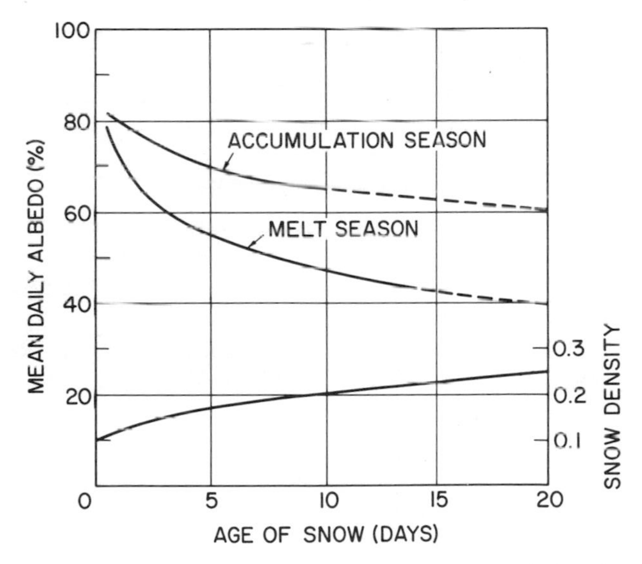

Fig. 3.3 Decrease of albedo and increase of the snow density with time. Drawn after US Army Corps of Engineers (1956).

following aspects refer to the melting process and thus to potential economic benefits of remote sensing:

The heat of fusion of ice is 79.7 cal g^{-1} which means that a heat gain of about 80 calories per 1 cm^2 melts a depth of 1 cm of water from pure ice at 0°C.

The thermal quality of snow is the ratio of the amount of heat required to produce a given volume of water from snow to the amount of heat required to melt the same volume of water from pure ice at 0°C. The thermal quality has been used (Wilson, 1941; US Army Corps of Engineers, 1956) to characterize the varying conditions of a snowpack. A snowpack with a water quality of 0.9 would require only 72 calories per 1 cm^2 in order to release 1 cm water depth. Such conditions occur if the snow is at 0°C and contains free water which is released when the ice matrix melts. If the temperature of snow is below 0°C, additional heat is required to raise the temperature to the melting point. Taking into account the specific heat of ice, which is about 0.5 cal g^{-1} $°C^{-1}$, this effect is relatively small: it needs only 4 calories to bring a grain of snow with an initial temperature of −8°C to 0°C so that the thermal quality of this cold snow would be 1.05.

The assessment of melting factors becomes more complicated in a continuous melting process: while the effect of negative temperatures with regard to the cooling of snow appears to be small, it is probably more important with regard to the refreezing of meltwater within the snowpack.

Snowmelt by warm rain can be calculated as follows:

$$M = P_r \frac{T_r}{80} \tag{3.5}$$

where M is the depth of water melted by rain in mm, P_r is the rainfall depth in mm, T_r is the temperature of rain in °C (which can be replaced by the wet bulb temperature) and 80 is the ratio of the heat of fusion of ice to the specific heat of water in °C (in terms of units, 80 = 80 cal g^{-1}/1 cal g^{-1} $°C^{-1}$).

For example, if P_r = 20 mm and T_r = + 8°C, the water depth melted by this rain is only 2 mm. The energy of impact of falling rain is negligible. However, as suggested by Anderson (1970), although rain does not melt much snow, considerable snowmelt may occur during rainfall, due to the absorption of latent heat from condensation on the snow cover surface.

The latent heat of sublimation of ice, if taken as a total of the latent heat of vaporization of water (596 cal g^{-1} at 0°C) and of the latent heat of fusion of ice, amounts to 676 cal g^{-1}. Ice and water require relatively large amounts of heat for a change of state. The snow cover is a good insulator of the soil in the winter. Large-scale remote sensing can be used to detect the absence of the seasonal snow cover in certain years for predictions of harvest failures.

3.3 Seasonal snow cover

In alpine countries with a wide altitude range and varying climatic conditions, the snow cover has a variable duration, as follows.

Temporary snow cover exists for several days and is formed in low altitudes during the winter, in high altitudes during the summer.

Seasonal snow cover is formed during weeks or months by consecutive snowfalls and then gradually disappears during the snowmelt season in a yearly cycle.

Above a certain altitude and in some localities, snow becomes firn or ice and is carried over from year to year.

Plate II shows all these types of snow cover in the typical alpine country of Switzerland as viewed by Landsat in March 1976. A small portion of the total snow covered area might have been only temporary on the day of overflight. The image shows the seasonal snow cover which disappeared in the next 4–5 months. Only 1342 km^2 (Bezinge and Kasser, 1979) or 3.25% of the surface area of Switzerland remains permanently covered by glacier ice and firn.

The duration of the seasonal snow cover is variable from year to year. Figure 3.4 shows maximum, minimum and average snow depths observed at Weissfluhjoch, 2540 m a.s.l., in the eastern Swiss Alps, in the past 40 years. Figure 3.5 illustrates the deposition of 12 consecutive snow layers during the build-up of the snow cover in the year 1973, at the same station.

As noted from Fig. 3.6, the seasonal snow cover disappears earlier in lower altitudes. The figure also shows different snow accumulation patterns in the respective years: in 1953, the snow cover built up gradually and reached the

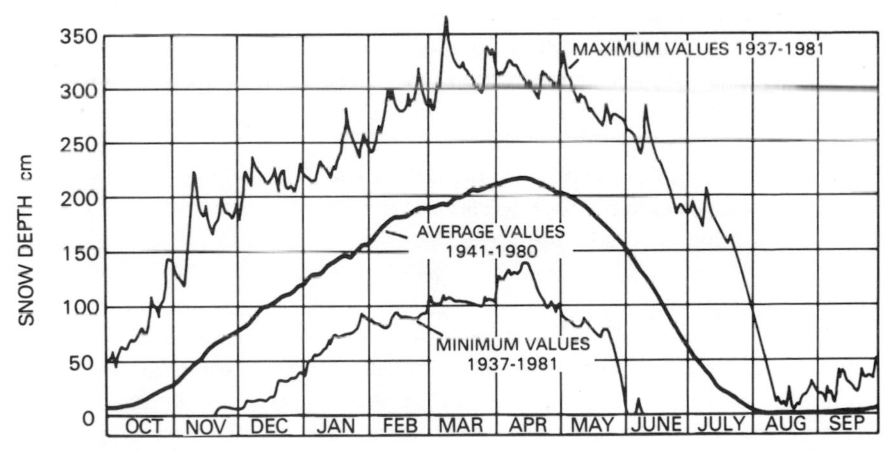

Fig. 3.4 Maximum, minimum and average snow depths as observed at Weissfluhjoch near Davos, 2540 m a.s.l., eastern Swiss Alps.

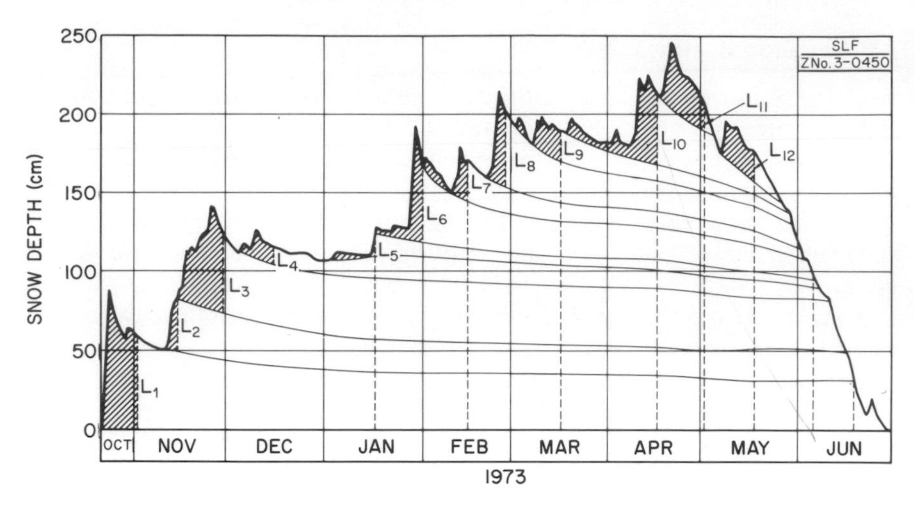

Fig. 3.5 Deposition of 12 consecutive snow layers during the build-up of the snow cover at Weissfluhjoch, 2540 m a.s.l., in 1973.

maximum water equivalent in February; in the hydrological year (October–September) of 1960, early snowfalls caused a maximum snow depth already in November. However, the corresponding water equivalent was relatively small due to the low density of the new snow. Recalling Equation (3.4), Fig. 3.6 demonstrates the role of time in the gradual increase in snow density.

Due to the temperature lapse rate, the maximum accumulation of snow is not synchronous at different altitudes. At Weissfluhjoch, 2540 m a.s.l., it is reached in April or May while in Disentis, 1173 m a.s.l., it generally occurs in February. It is therefore difficult to set even an average single date of maximum accumulation for basins with a considerable elevation range. While the snowmelt season already starts in the lower parts, the accumulation of snow continues at the higher altitudes. This situation is typical for mountain basins in which snow is a major runoff factor.

When the accumulation and ablation of the seasonal snow cover is monitored by satellites, it is recommended to interpret the data not for a basin as a whole, but separately for several elevation zones.

3.4 Snow-cover mapping

The snow cover melts mainly from the surface, apart from some melting from the ground under certain conditions. Although the exposure to the atmosphere and especially to the reflected radiation is increased for scattered snow patches in comparison with a compact snow cover, the volume of meltwater production is for a given input approximately directly proportional to the snow-covered

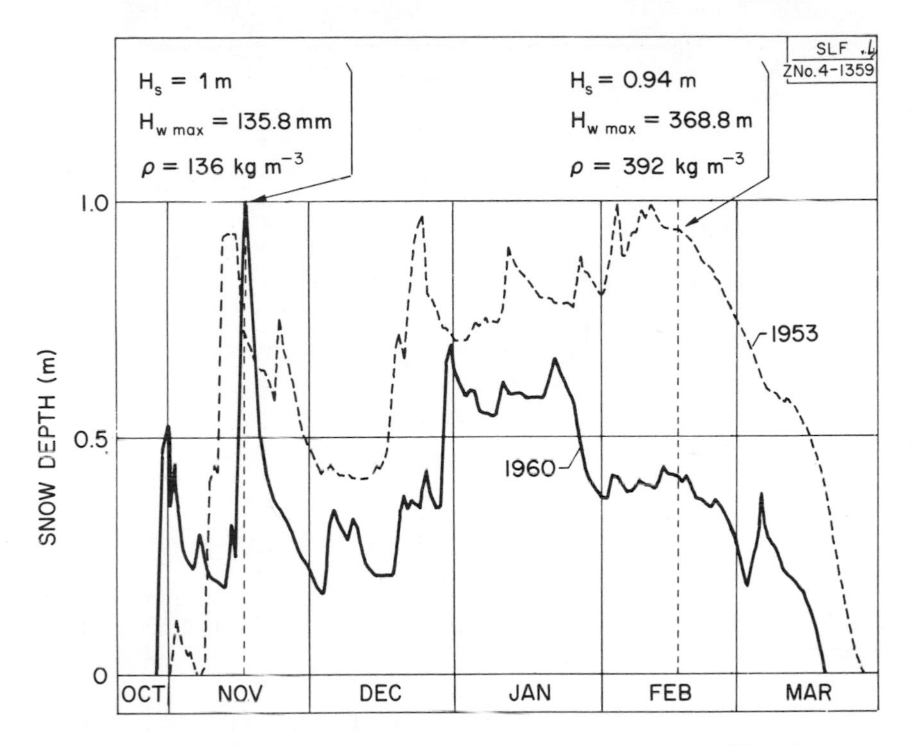

Fig. 3.6 Accumulation and depletion of the seasonal snow cover at the Disentis station, 1173 m a.s.l., eastern Swiss Alps.

area. The energy input is frequently represented by the degree-day factor and the meltwater volume is calculated as follows:

$$V_m = aTA \tag{3.6}$$

where V_m is the meltwater volume (m^3), a is the degree-day factor (m °C^{-1} d^{-1}), T is the number of degree-days (°C d) and A is the snow-covered area in m^2.

Collins (1934) introduced the degree-days to runoff predictions by relating accumulated degree-days to accumulated runoff depths as shown in Fig. 3.7. In this relation, the effect of temperature on snowmelt is distorted in two ways:

(1) The gradual decrease of the snow-covered area is disregarded.
(2) The transformation of the meltwater input by the basin is included.

Linsley (1943) took into account the changing snow-covered area by referring to the position of the snowline and derived degree-day ratios which increased during the snowmelt season. His high values for July (exceeding 0.7 cm °C^{-1} d^{-1} might also have been influenced by the basin response, in particular by the effect of the recession flow.

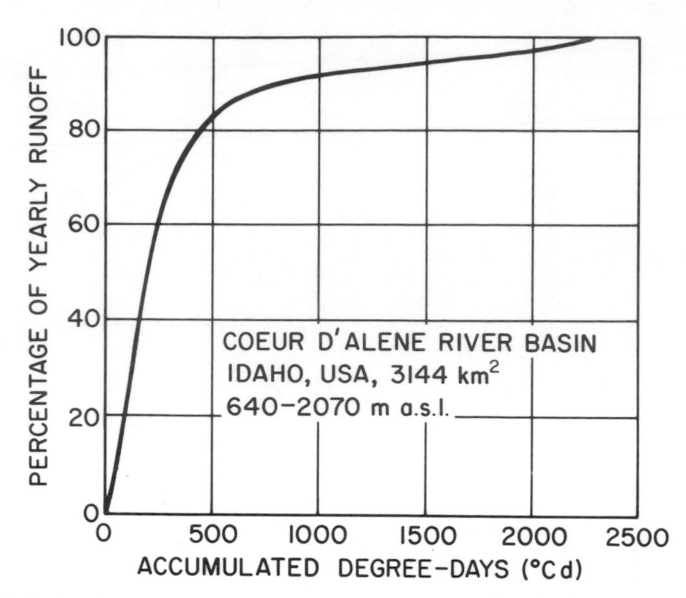

Fig. 3.7 Relation between accumulated degree-days and accumulated runoff depths. Figure drawn after Collins (1934).

Even nowadays, the effect of the snow-covered area is sometimes included in degree-day ratios which leads to areal values differing from point values. This confusion is easily eliminated by using the point values of the degree-day factor and by taking into account the snow coverage of a basin in calculating snowmelt.

Early attempts to forecast runoff from the areal extent of the snow cover used panoramic terrestrial photographs (Potts, 1937). In this study the deviations from a relation between the snow covered area and the volume of the subsequent runoff in the respective years were attributed to the varying precipitation in the snowmelt season (Potts, 1944). However, the main reason is that there is no unequivocal relation between the snow-covered area and the stored water volume. Figure 3.8 shows terrestrial photographs of an alpine snow cover on two different dates. The snow covered area looks similar in both cases but the water equivalent measured at 2540 m a.s.l. was 410 mm on 16 May, 1969 and 800 mm on 22 June, 1970. Thus, the accumulation of snow varied roughly in a ratio 1:2. The difference between these two snowpacks would have been revealed by sequential photographs. With the same energy input, the snow cover of 1969 would have disappeared more rapidly than that of 1970.

It is thus evident that snow-cover mapping, in contrast for example to geological mapping, must be frequently repeated in order to be useful for the assessment of the water storage. This requirement means more working hours and expenses for terrestrial and aircraft mapping but it is effortlessly fulfilled by satellites.

(a)

(b)

Fig. 3.8 Terrestrial photographs of alpine snow cover in the area Weissfluhjoch-Dischma; (a) 16 May, 1969; (b) 22 June, 1970.

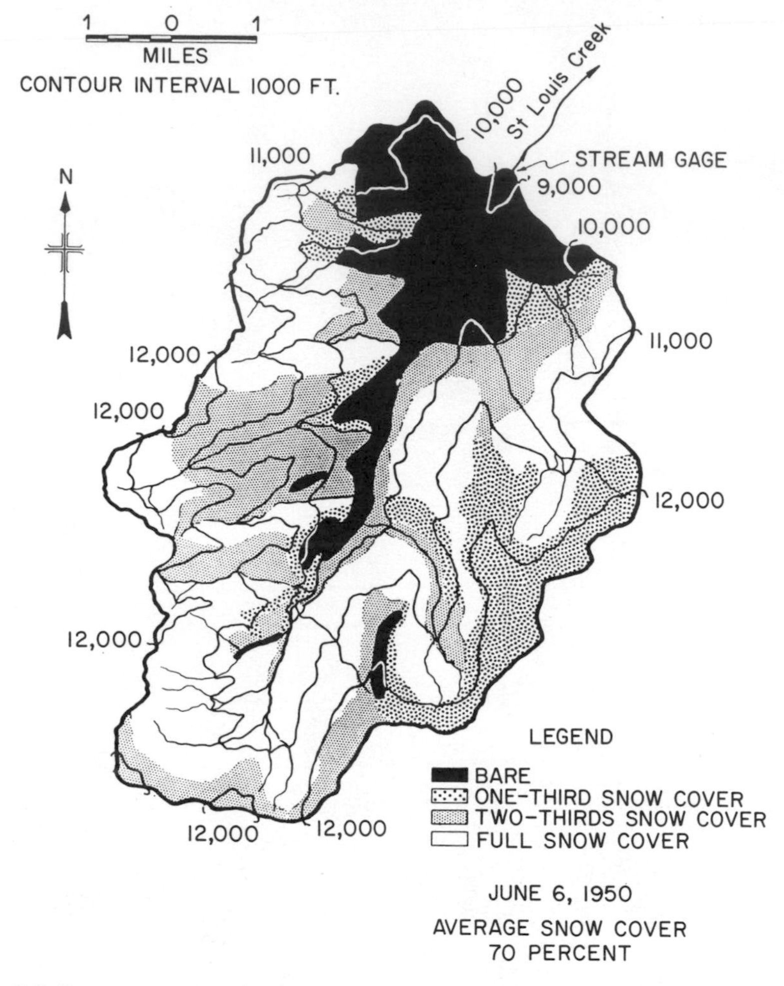

Fig. 3.9 Snow-cover mapping by terrestrial observations in the St Louis Creek basin, Colorado, USA, situation on 13 June, 1950. (Reproduced from Garstka *et al.*, 1958.)

The need for snow mapping was recognized long before the advent of remote sensing. In the experimental basin of St Louis Creek in the Rocky Mountains (Garstka *et al.*, 1958) the total area of 93 km^2 was divided into 16 topographic compartments in which the snow coverage was observed and evaluated. One such evaluation is shown in Fig. 3.9. Since this procedure must be repeated in not-too-long intervals, it is laborious and time consuming.

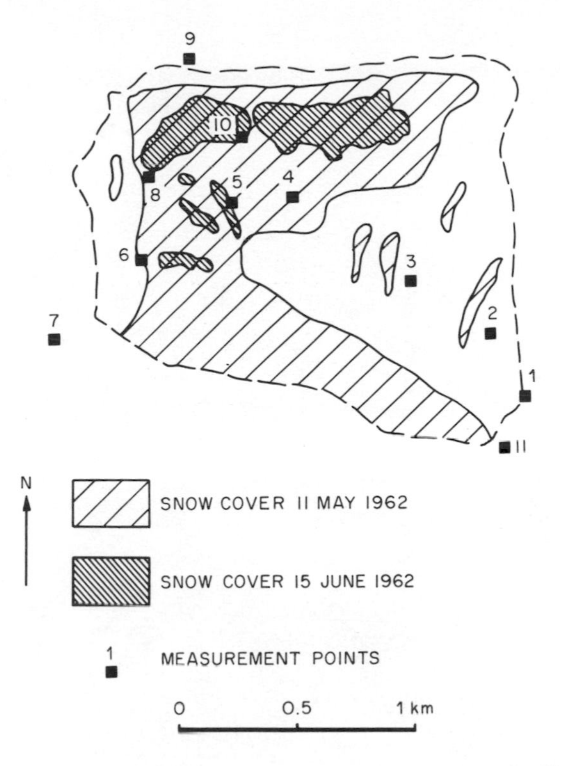

Fig. 3.10 Snow-cover mapping by terrestrial observations in the Modry Dul basin, Central Europe.

Terrestrial snow mapping is more expedient in very small basins which can be overlooked from a single point. Figure 3.10 shows the snow cover in the Modry Dul basin (2.65 km, 1000–1554 m a.s.l.) in Central Europe on two different dates. The terrain is not very rugged so that the snowline can be drawn without much difficulty.

Aerial photography provides more efficient and accurate means of periodically determining the snow-covered area. Figure 3.11 shows the snow cover in the Dischma representative basin in the Swiss Alps (43.3 km^2, 1668–3146 m a.s.l.) in different stages of the snowmelt season. Distortions by the central projection of the photographic camera have been rectified so that the original corresponds to a chart in a scale 1:50 000. The area outside the basin is covered by a mask which can thus be used for all orthophotographs produced. As mentioned before, the snow coverage must be evaluated separately for several elevation zones. This is done again by masking techniques. The masks for three elevation zones in the Dischma basin are shown in Fig. 3.12.

A hypothetical example in Table 3.3 illustrates the serious errors which may

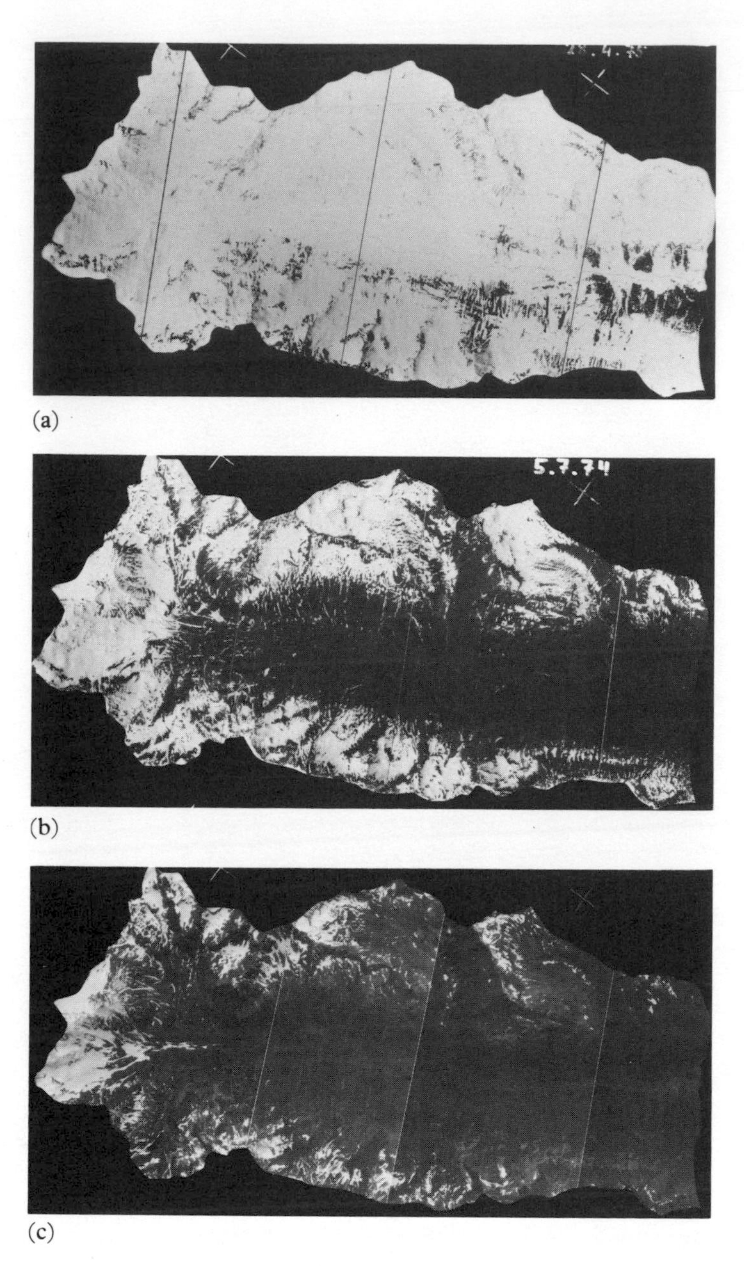

Fig. 3.11 Orthophotographs of the snow cover in the Dischma basin at different stages of the snowmelt season: (a) 28 April, 1975; (b) 5 July, 1974; (c) 15 August, 1978. Photograph: Swiss Air Force.

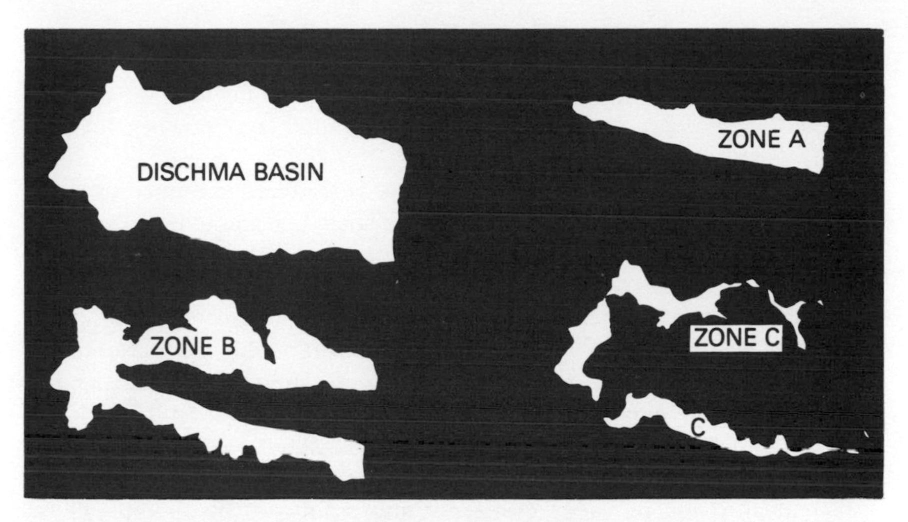

Fig. 3.12 Masks for the Dischma basin and for the elevation zones A, B, C. Zone A: 1668–2100 m a.s.l.; zone B: 2100–2600 m a.s.l.; zone C: 2600–3146 m a.s.l.

result if the temperature and the areal extent of the snow cover are taken only as an average for the whole basin. If the snowline is set at 50% snow coverage, the area below this line is interpreted as snow free (I) and the area above this line as totally snow covered (II). This gives a lower meltwater volume than the total of the partial meltwater volumes from three elevation bands. If an average value of the snow coverage is used and the meltwater production is calculated for the

Table 3.3 Meltwater production from different assessments of the snow coverage

Zone	*Area* ($\times 10^6 m^2$)	*Snow coverage*	*Degree-days* (°C d)	*Degree-day factor* ($cm\,°C^{-1}\,d^{-1}$)	*Meltwater production* (m^3)
A	10	0.25	7.5	0.5	93 750
B	20	0.5	4.5	0.5	225 000
C	10	0.75	1.5	0.5	56 250
Total	40				375 000
I	20	0	6.75	0.5	0
II	20	1.0	2.25	0.5	225 000
Total	40	0.5	4.5	0.5	225 000
Entire basin	40	0.5	4.5	0.5	450 000

whole basin by the average number of degree-days, the volume obtained is too high.

This example was computed for a difference of 500 m between the average elevations of the zones A, B, C and for a temperature lapse rate of 0.6°C per 100 m.

Since the snow cover in the European Alps is scattered in numerous snow patches, it is practically impossible to draw a snow line and to determine the snow-covered area by planimetering. Therefore, a transparent orthophotograph is projected onto a screen and a Quantimet 720 computer is used to compute about 500 000 points which appear either as snow covered or as snow free (Martinec, 1973). Since about 200 000 points fall within the boundaries of the Dischma basin, each point represents an area of 270 m^2, so that the spatial resolution of this evaluation is about 15 m. Thus, the resolution is not determined by the original orthophotograph, in which case it amounts to 1–5 m, but by the method of evaluation.

The advance and retreat of glaciers is also monitored by aircraft. These changes are naturally much slower than those of the seasonal snow cover. Photographs are usually taken in the autumn to see the difference from the previous year. As an example, a photograph of a small glacier is shown in Fig. 3.13.

Coming back to the snow cover, a variety of artificial satellites can be used more efficiently and on a larger scale than aircraft for mapping. Various possibilities are summarized in Table 3.4 (Lenco, 1982; Rango *et al.*, 1984).

A detailed description of platforms and sensors is given in Chapter 2. The repeat period for Landsat has been improved recently by introducing a 'jumping orbit'. This means that when a basin was previously seen on two consecutive overflight days because of overlapping coverage, it is now seen on the first and ninth days. In addition, the overflight interval has been shortened to 16 days with the launch of Landsat 4. The spatial resolution of the NOAA satellites is about 1.1 km near nadir and deteriorates towards the boundaries of an image.

As indicated in Table 3.4, of the many sensors and platforms available, only Landsat has a sufficient resolution to be used in small basins. Actually, the minimum size should refer to the area of the smallest partial zone of a basin in which the snow coverage is evaluated, rather than to the total area of a basin. Therefore, the use of satellites other than Landsat is limited to larger basins. On the other hand, snow-cover monitoring by Landsat is hampered in certain regions by a frequent cloud cover. For the NOAA satellites, the chance of obtaining cloud-free images is much better because the frequency of coverage is greater. With the geostationary satellites, snow-cover data can be obtained at practically any desired time.

In addition to the areal extent of snow cover, regional information on the water equivalent of snow and on the onset of snowmelt can be obtained by the

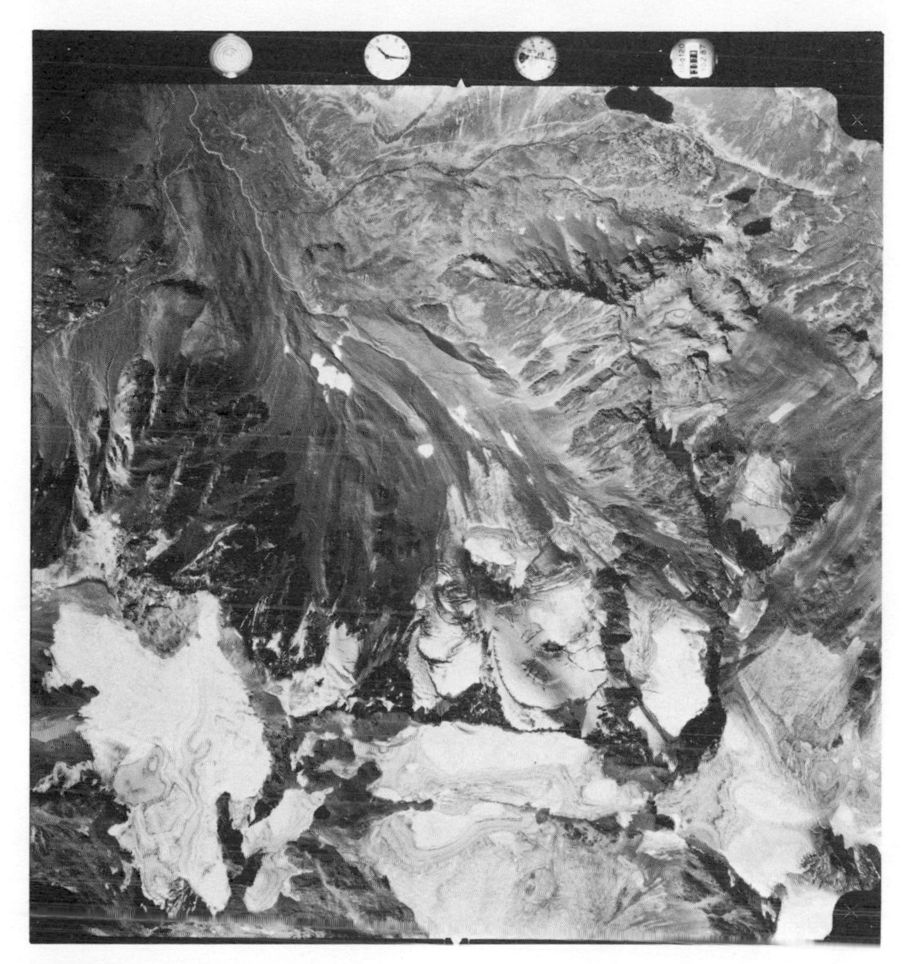

Fig. 3.13 Aircraft monitoring of glaciers: Scalletta glacier in the Dischma basin on 13 September, 1973. Photograph of the Federal Office for National Topography, Bern, Switzerland.

microwave radiometers on the Nimbus satellites (Kunzi *et al.*, 1982). The interpretation of these large-scale data is still under development. Efforts are in progress to use the microwave emission from snow for measuring the depth and water equivalent of the snow cover (Foster *et al.*, 1984).

Errors can arise from the differences in emissivity of various surfaces, variations in atmospheric attenuation and differences in elevations of terrain (Barnes and Bowley, 1974). In this connection, it should be mentioned that the natural radioactivity of the Earth offers another possibility for measuring the water equivalent of the snow cover on an areal basis. The gamma-radiation,

Table 3.4 Possibilities of remote sensing for snow-cover mapping

Platform sensor	*Spatial resolution*	*Minimum basin size digital (photo)*	*Repeat period*
Aircraft			
Orthophoto	3 m	1 km^2	flexible
Landsat			
MSS	57 × 79 m	10/20 km^2	16 days
RBV	40 m	5/10 km^2	18 days
TM	28.5 m	2.5/5 km^2	16 days
NOAA			
AVHRR	1.1 km	200/500 km^2	12 h
GOES			
VISSR	1.1 km	200/500 km^2	30 min
Meteosat			
Visible	2.5 km	500/1000 km^2	30 min
Near infrared	5 km		
Satellite HCMM (1978–1980)			
Visible and near infrared,	500 m	100/250 km^2	1.5–3.5 days
Nimbus-7			
Visible, near infrared, thermic infrared	800 m	200/500 km^2	1–6 days
SMMR	20–100 km		

MSS multiple spectral scanner; RBV return beam vidicon; TM thematic mapper; NOAA National Oceanic and Atmospheric Administration; GOES geostationary operational environmental satellite; VISSR visible and infrared spin scan radiometer; HCMM heat capacity mapping mission; SMMR scanning multichannel microwave radiometer.

which is attenuated in relation to the water equivalent of the snow cover, is detected from an aeroplane. Such measurements have been reported from Scandinavia (Dahl and Ødegaard, 1970), the USA (Peck *et al.*, 1971) and USSR (Dmitriev *et al.*, 1973). Effects of soil moisture, background radiation and air density must be taken into account in order to avoid errors. The method is preferably used in flat areas. Most authors give 30–35 cm as the maximum water equivalent which can be reliably measured.

Recent experiments with a synthetic aperture radar (Matzler and Schanda, 1983) indicate possibilities of snow-cover monitoring even through cloud cover and with an improved spatial resolution. Problems still to be overcome concern the interpretation of large amounts of incoming data and large distortions of the mountainous terrain if the sensor is mounted on an aeroplane.

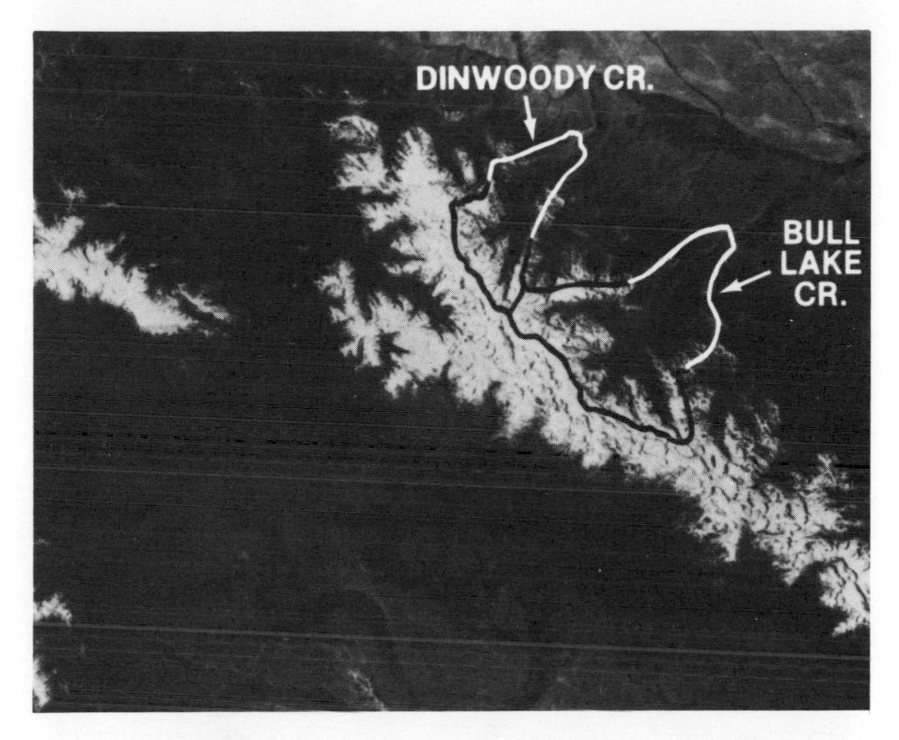

Fig. 3.14 Landsat image of the snow cover in Wyoming, USA, 28 June, 1976 with the boundaries of the Dinwoody Creek basin and Bull Lake Creek basin (Rango, 1980).

The chance of cloud-free conditions is frequently better at night than during the day. The light of the full moon appears to be sufficiently bright for snow-cover mapping in the visible range by the DMSP satellite, but not with the present sensors of Landsat (Foster, 1983).

While efforts to ensure all-weather remote sensing and additional information on the snow cover are in progress, the monitoring of snow-covered areas is already operational. There are two main ways of processing the data:

(1) By using a zoom transfer scope, a satellite image is superimposed on a map and the snowline is drawn onto the map. The snow-covered area can be determined by planimetering. This method is possible in areas or situations in which the snow cover is not too scattered so that a snowline can be determined. Figure 3.14 shows a Landsat image of the Dinwoody Creek basin (228 km^2, 1981–4202 m a.s.l.) and Bull Lake Creek basin (484 km^2, 1790–4202 m a.s.l.) in Wyoming, USA (Rango, 1980). The more-or-less rounded surface of the Rocky Mountains allows such evaluation. An advantage of this method is the possibility of drawing a snowline even if the snow cover is obscured by forest canopy.

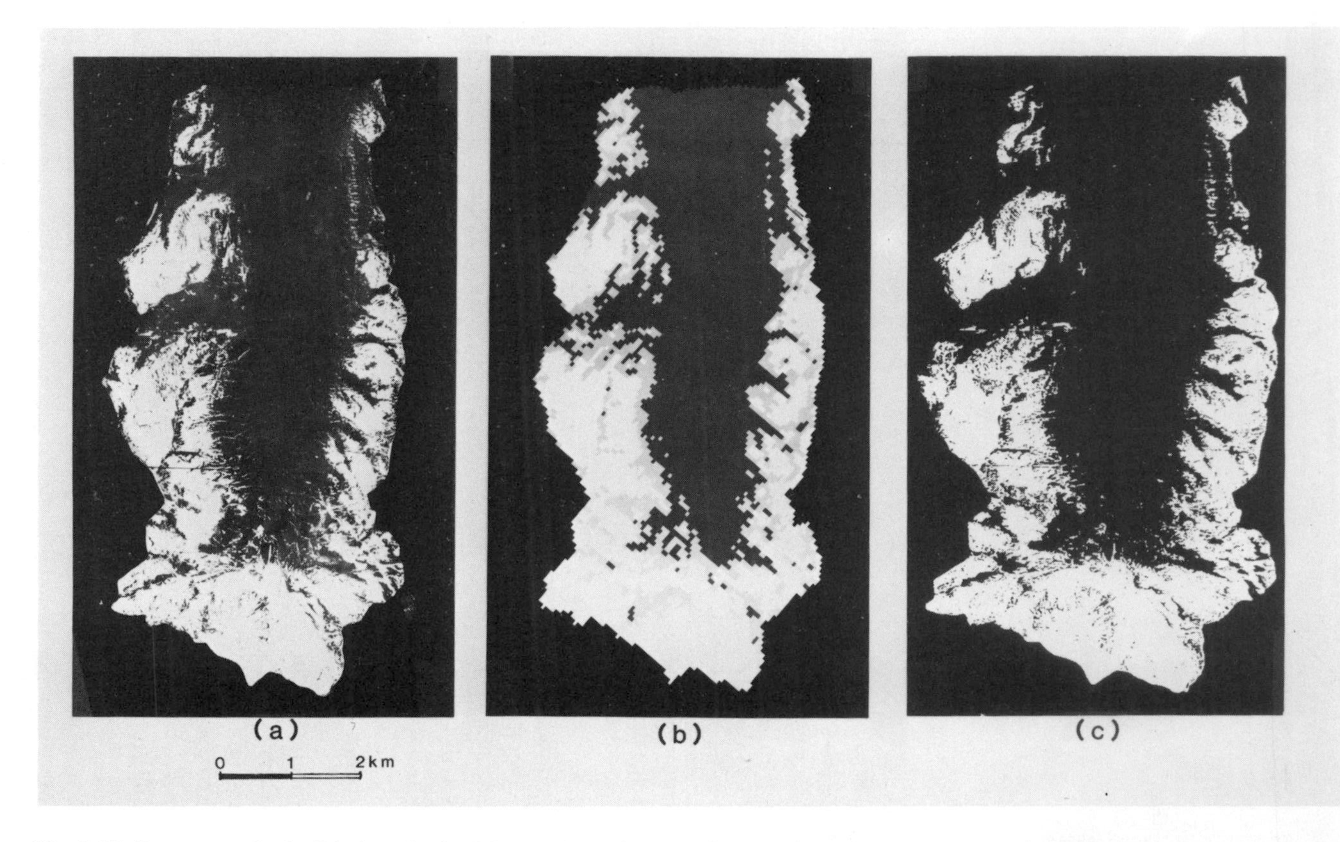

Fig. 3.15 Snow cover in the Dischma basin, 8 June, 1976, displayed from (a) aircraft orthophotograph (Photograph: Swiss Air Force); (b) Landsat MSS classified image; (c) computer printout from orthophotograph after edge-enhancement preprocessing.

(2) Digital processing of data stored on the magnetic tape is necessary in rugged terrain where a usable snowline hardly exists.

The snow-covered area is obtained by counting the number of snow-covered and snow-free pixels (a Landsat pixel is approximately 60 × 80 m). Figure 3.15 (a) and (b) shows a comparison of this evaluation with an orthophotograph of the Dischma basin in the Swiss Alps. The digital data have been converted into an image by Photomation Optronics. The pixels are classified into three categories: (1) snow-covered = white; (2) partially snow-covered = gray; (3) snow-free = dark. Pixels outside the boundaries of the basin are disregarded. Figure 3.15(c) is a computer printout from the orthophotograph in Fig. 3.15(a) which has a resolution of about 10 m after edge enhancement preprocessing (Good, 1983). The width distortion is caused by the horizontal and vertical spacing of the printer. Again, the snow-covered and snow-free points are counted in order to obtain the snow coverage. The main purpose of this processing of orthophotographs is to study the patterns of snow fields in an attempt to estimate not only the areal extent but also the volume of snow from two-dimensional orthophotographs. Figure 3.16 shows a Landsat image of the Dischma basin in a later stage of the snowmelt season.

Various aspects of Landsat data processing including the selection of the available sensor frequencies are dealt with in the literature (for example,

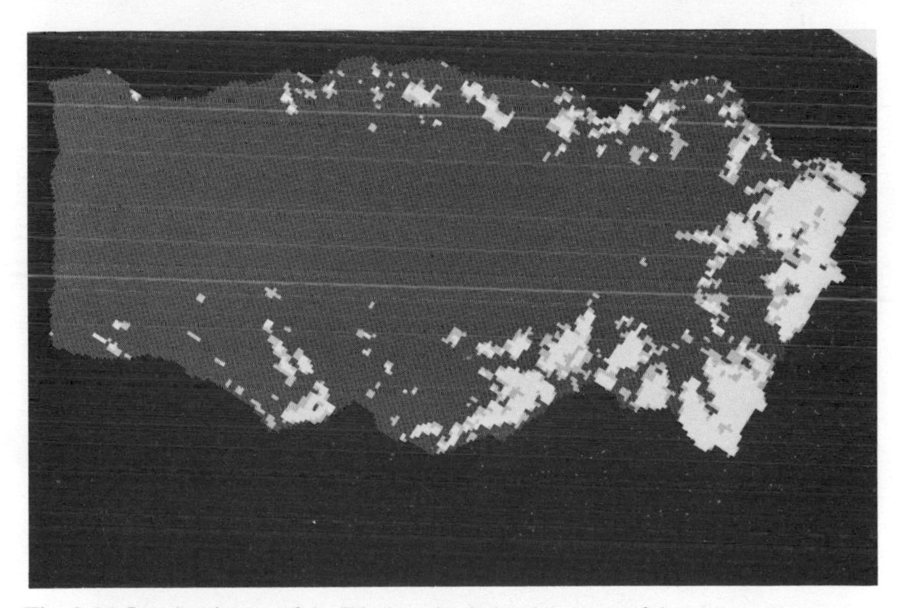

Fig. 3.16 Landsat image of the Dischma basin in a late stage of the snowmelt season on 18 July, 1979. Published by courtesy of the Institute of Geography, University of Zurich, and Institute for Communications Techniques, Institute of Technology, Zurich, Switzerland.

Haefner and Seidel, 1974; Rango and Itten, 1976; Haefner, 1980). In particular, a special channel on 1.55–1.75 μm wavelength has been introduced in order to distinguish between clouds and snow. Thus, the clouds can be better identified, but unfortunately still not seen through by the Landsat sensors.

As an example of large-scale snow-cover mapping, Plate III shows the snow cover in Europe as viewed by NOAA–6.

The NOAA data can be evaluated by photointerpretation or by digital analysis. An alternative digital approach is described by Andersen (1982): the reflectance in areas of known snow coverage is compared with the reflectance of pixels covering an entire drainage basin. In areas of known snow coverage, the reflectance range between snow-free and snow-covered pixels is calculated. A linear relation between the pixel reflectance or brightness level and the snow coverage is then derived and used to calculate the snow coverage over the entire basin. Another example of the NOAA imagery is shown in Fig. 3.17. An

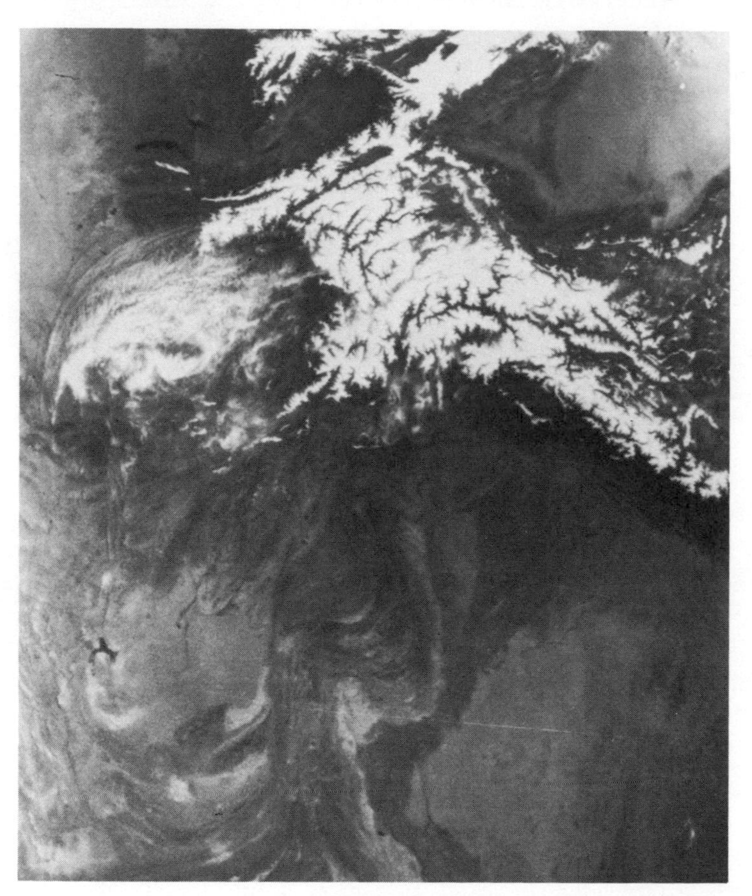

Fig. 3.17 NOAA-2 image of the snow cover in the Himalayas on 4 April, 1975.

example of snow-cover mapping by Nimbus in the Austrian Alps is given in Plate IV.

Large-scale snow-cover mapping in terms of continents and hemispheres is carried out mainly with regard to study of the global climate and is described in the pertinent literature (for example, Foster *et al.*, 1983; Deutsch *et al.*, 1981; Kukla *et al.*, 1981; Matson and Wiesnet, 1981). For hydrological applications, the following aspects of snow-cover mapping are important:

(1) Adequate resolution for the given basin size
(2) Periodical evaluations during the snowmelt season
(3) Distinguishing between seasonal and temporary snow cover.

These items are dealt with in the next section.

3.5 Snow-cover depletion curves

Depletion curves of snow-covered areas continuously indicate the gradual areal diminishment of the seasonal snow cover during the snowmelt season. As seen in Fig. 3.18, they are derived by connecting points of a known snow coverage as determined by the methods described. A long interval between the measurements increases the uncertainty and risk of errors in drawing these curves. In Fig. 3.18 separate curves are drawn for the respective elevation zones of the alpine basin of Dischma:

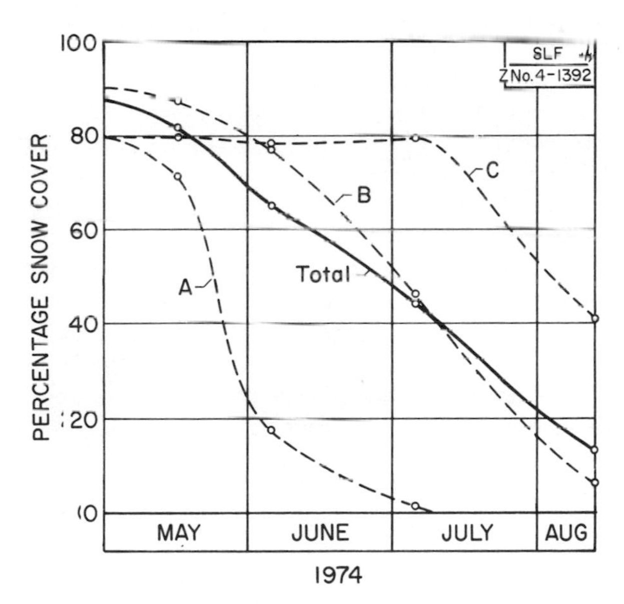

Fig. 3.18 Depletion curves of snow-covered areas in the alpine basin Dischma, eastern Swiss Alps, for the total basin and for the elevation zones A, B, C.

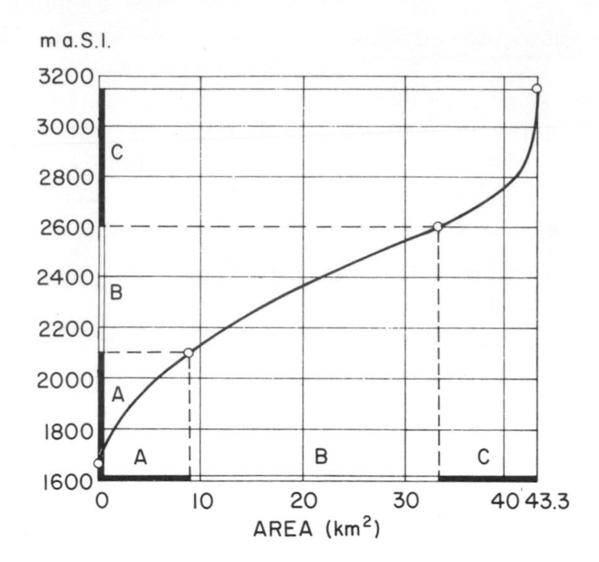

Fig. 3.19 Area–elevation curve and partial areas of the Dischma basin.

Zone A, 1668–2100 m a.s.l., 8.9 km^2
Zone B, 2100–2600 m a.s.l., 24.5 km^2
Zone C, 2600–3146 m a.s.l., 9.9 km^2

The curve for the total basin is an average of the zone curves weighted by the respective areas. These areas are obtained from the area–elevation curve of the basin which is shown in Fig. 3.19. In the given case, the altitude range of the basin has been divided into three roughly equal parts. The changing snow coverage can be read off these curves each day and used for calculating the daily meltwater production in each zone. The minimum area required for a given spatial resolution of a sensor thus refers to the smallest elevation zone of a basin.

Naturally, the snow cover must be monitored again in each new year because the depletion curves always take a different course depending on the initial snow reserves, meteorological conditions and intermittent precipitation during the snowmelt season. Figure 3.20 shows the total depletion curve from Fig. 3.19 as compared with such curves for two other years in the Dischma basin. The deepest snowpack was in 1970, but the depletion curve declined rapidly in June due to high temperatures.

The typical shape of depletion curves can be approximated by the equation (Leaf, 1967):

$$A = \frac{100}{1 + \exp(-bt)} \tag{3.7}$$

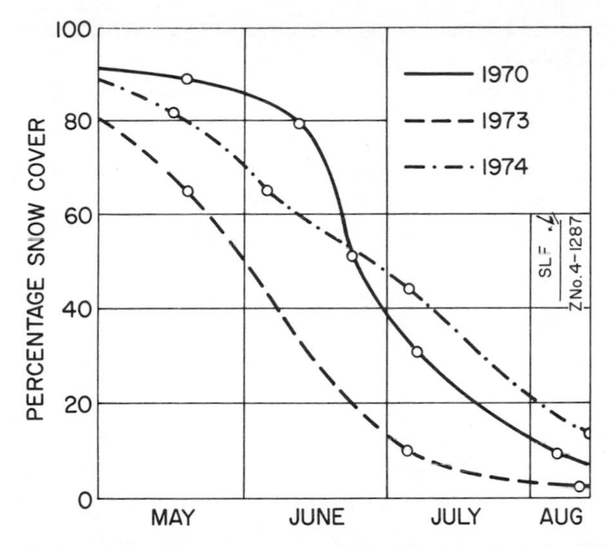

Fig. 3.20 Depletion curves of snow coverage in the Dischma basin in different years.

where A is the percentage of area without snow, t is the time measured from an arbitrary origin and b is a coefficient. The equation can be rearranged as:

$$S = \frac{100}{1 + \exp(bn)} \tag{3.8}$$

where S is the snow-covered area in %, b is a coefficient and n is the number of days before (−) or after (+) the date at which $S = 50\%$.

Referring to the depletion curve for 1970 in Fig. 3.20, S_{50} is reached on 20 June when $n = 0$:

$$S = \frac{100}{1+1} = 50\%$$

On 1 May, $n = -51$. By substituting $b = 0.05$:

$$S = \frac{100}{1 + \exp(0.05 \times (-51))} = \frac{100}{1078} = 92.8\%$$

On 10 August, with $n = +51$:

$$S = \frac{100}{1 + \exp(0.05 \times 51)} = \frac{100}{13\,807} = 7.2\%$$

In 1974 the decline of the depletion curve is less steep. This could be roughly taken into account by reducing the value of b to 0.04. Even so, it is difficult to predict the course of a depletion curve in a current year by Equation (3.8) or by other, more complicated formulas.

The gradual decrease in the snow coverage is typical for the alpine snow cover and has two reasons:

(1) The irregular deposition of snow which results, even on a plain, in variable snow depths. The subsequent melting of snow layers leads to the disappearance of snow in a gradually increasing part of the total area as is illustrated in Fig. 3.21.

(2) Due to the temperature lapse rate, the snowmelt is progressing from the lower parts of a basin to the upper parts. The characteristic form of the area–elevation curve as shown in Fig. 3.19 is reflected in the S-shaped depletion curves of the snow coverage.

Figure 3.22 demonstrates the practical importance of this phenomenon for river flow. The lysimeter represents a small area of 5 m^2 which is completely covered with snow during practically the whole snowmelt season, then becomes snow free in 2 or 3 days. Consequently, the runoff from such a lysimeter rises in accordance with the temperature (degree-days) and then ceases abruptly. In contrast to this hydrograph, snowmelt runoff from a basin is governed by the depletion curve of the snow cover. The meltwater production is already on the

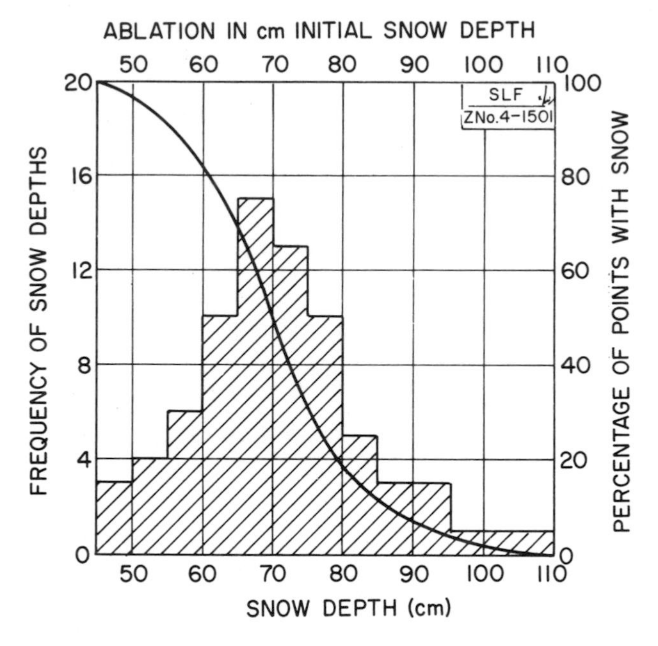

Fig. 3.21 Frequency distribution of snow depths as measured on a snow course and the resulting gradual decrease of the snow-covered area by melting.

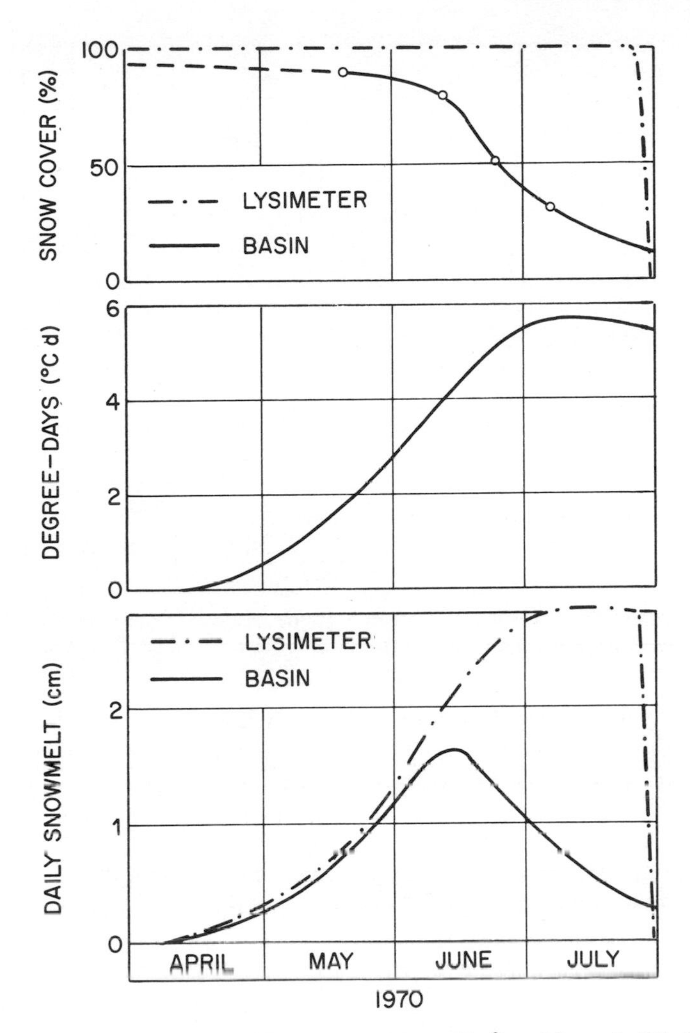

Fig. 3.22 Runoff hydrographs from a snow lysimeter (5m^2) and from the Dischma basin (43.3 km^2) as a result of temperatures and snow-covered areas.

decline while the temperature is still rising. This phenomenon reduces the risk of floods from snowmelt.

In the rugged terrain represented by the Dischma basin, 100% snow coverage is never reached. The value of 90% in Fig. 3.20 does not mean that 90% of the area is above a snowline. It means that sharp and steep rocks at all elevations (especially at high elevations) are not covered with snow. Consequently, the decrease in snow coverage progresses in all elevation zones, even if with a certain time shift. A different basin character is reflected by the depletion curves in Fig. 3.23. In the more rounded terrain of the Wind River Mountains in Wyoming,

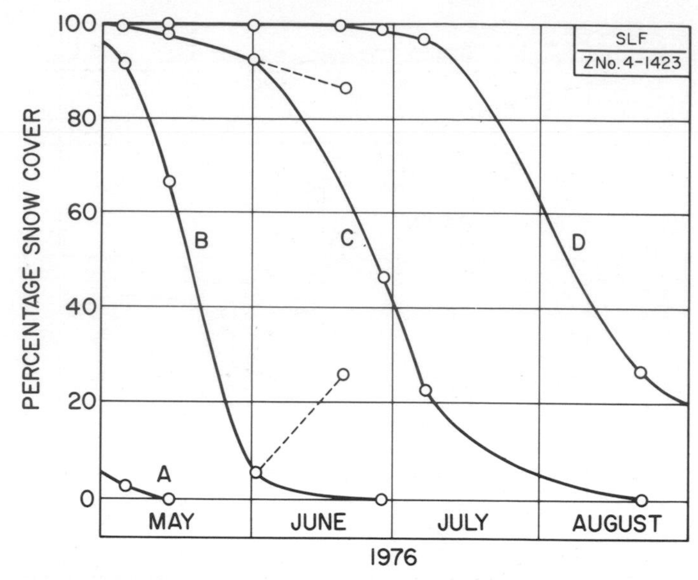

Fig. 3.23 Depletion curves of snow coverage in the Dinwoody Creek basin, Wyoming, USA. Points on 19 June refer to temporary snow cover following a snow storm.

USA, the top zone D is still 100% snow covered while the two lowest zones A and B are already snow free. Figure 3.23 also illustrates a risk of misinterpreting the satellite data in deriving the depletion curves: on 19 June 1976, a snow storm just preceding a Landsat overflight optically increased the snow cover to 86% in zone C and to 26% in zone B. A daily monitoring of the snow cover would have shown that this thin snow cover disappeared in a matter of hours or days. With an overflight interval of 18 days, the short-lived snow cover as seen on 19 June 1976 might have affected the drawing of depletion curves well before and after this date.

The risk of erroneous drawing of depletion curves due to summer snow storms is further increased if the interval between usable Landsat images becomes too long. A hypothetical example in Fig. 3.24 shows a temporary increase in the snow coverage from 50% to 100% by a snowfall in early June. If the preceding Landsat overflight did not deliver usable data, for example because of clouds, the points from overflights 2, 4 and 5 might have been connected by a false depletion curve. As a result, an excessive meltwater production would have been calculated. Such errors could discredit remote sensing as a tool for improving the reliability of river-flow forecasts.

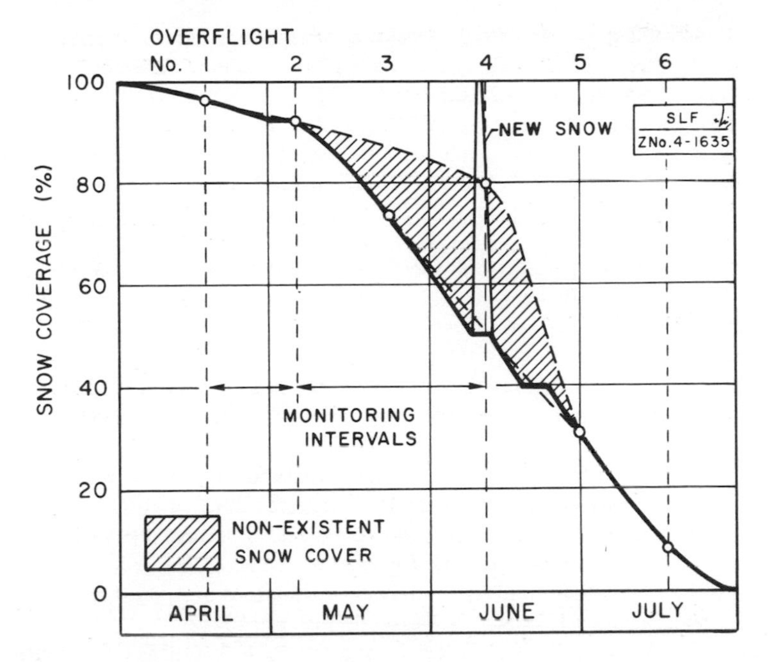

Fig. 3.24 Hypothetical example of possible distortion of a depletion curve due to a temporary increase in the snow coverage by a summer snowfall and to missing Landsat data from the preceding overflight.

References

Andersen, T. (1982) Operational snow mapping by satellites, *Hydrological Aspects of Alpine and High-Mountain Areas*, Proceedings Exeter Symposium, IAHS Publication No. 138, pp. 149–54.

Anderson, H.W. (1970) Storage and delivery of rainfall and snowmelt water as related to forest environments, *Proceedings 3rd Forest Microclimate Symposium, Seebe, Alberta 1969*, Canadian Forestry Service, Calgary, Alberta, pp. 51–67.

Bader, H. (1962) *The Physics and Mechanics of Snow as a Material*, Cold Regions Research and Engineering Laboratory, Hanover, NH, Report II-B, p. 1.

Barnes, J.C. and Bowley, C.J. (1974) *Handbook of Techniques for Satellite Snow Mapping*, prepared for NASA/Goddard Space Flight Center, Greenbelt, MD, ERT Document No. 0407-A.

Baumgartner, A. and Reichel, E. (1975) *Die Weltwasserbilanz* (World water balance), Oldenbourg, Munich.

Bezinge, A. and Kasser, P. (1979) Gletscher and Kraftwerke (Glaciers and power plants). In *Schweiz und ihre Gletscher*, Kümmerly and Frey Geographischer Verlag Bern, pp. 166–83.

Collins, E.H. (1934) Relationship of degree-days above freezing to runoff. *Trans. Am. Geophys. Union*, pp. 624–9.

Dahl, J.B. and Ødegaard, H. (1970) Areal measurement of water equivalent of snow deposits by means of natural radioactivity in the ground, *Symposium on Isotope Hydrology*, International Atomic Energy Agency, Vienna, pp. 191–210.

Deutsch, M., Wiesnet, D.R. and Rango, A. (eds) (1981) *Satellite Hydrology*, American Water Resources Association, Technical Publication Series TPS 81–1, pp. 157–236.

Diamond, M. and Lowry, W.P. (1953) *Correlation of Density of New Snow with 700 mb Temperature*, US Snow, Ice and Permafrost Research Establishment, Research Paper 1.

Dmitriev, A.V., Kogan, R.M., Nikiforov, M.V. and Fridman, Sh. D. (1973) The experience and practical use of aircraft gamma-ray survey of snow cover in the USSR, *Unesco/WMO/IAHS Symposia Banff 1972*, IAHS Publication No. 107, Vol. 1, pp. 702–12.

Donaldson, P.B. (1978) Melting of antarctic icebergs. *Nature (London)* **275**, 305–6.

Foster, J.L. (1983) Night-time observations of snow using visible imagery. *Int. J. Remote Sensing*, **4** (4), 785–91.

Foster, J., Owe, M. and Rango, A. (1983) Snow cover and temperature relationships in North America and Eurasia. *J. Climate Appl. Meteorol.* **22** (3), 460–9.

Foster, J., Hall, D.K., Chang, A.T.C. and Rango, A. (1984) An overview of passive microwave snow research and results. *Rev. Geophys. Space Phys.*, **22**, 195–208.

Garstka, W.U. (1964) Snow and snow survey. In *Handbook of Applied Hydrology* (ed. V.T. Chow), McGraw-Hill, New York, section 10, pp. 10–12.

Garstka, W.U., Love, L.D., Goodell, B.C. and Bertle, F.A. (1958) Factors affecting snowmelt and streamflow, Fraser Experimental Forest, US Government Printing Office, Washington DC, p. 51.

GHO (1982) *Glossaire des Termes Hydrologiques avec Définitions* (Glossary of hydrological terminology with definitions), Groupe de travail pour l'hydrologie opérationnelle (GHO), Berne, Service hydrologique national.

Good, W. (1983) *Estimation par des méthodes de traitement d'images de la quantité d' eau stockée dans un bassin versant* (Estimation of the water quantity stored in a basin by the methods of image processing), Federal Institute for Snow and Avalanche Research, Weissfluhjoch/Davos, Switzerland, Internal Report.

Haefner, H. (1980) Digital mapping of mountain snow cover under European conditions. In *Operational Applications of Satellite Snowcover Observations*, Workshop, Sparks, Nevada, 1979, NASA Conference Publication 2116, pp. 73–91.

Haefner, H. and Seidel, K. (1974) Methodological aspects and regional examples of snow cover mapping from ERTS-1 and EREP imagery of the Swiss Alps, *Proceedings Symposium of Frascati 1974*, ESRO SP-100, pp. 155–65.

Hoinkes, H. (1967) *Glaciology in the International Hydrological Decade*, IUGG General Assembly, Bern, IAHS Commission on Snow and Ice, Reports and Discussions, IAHS Publication No. 79, pp. 7–16.

Kotlyakov, V.M. (1970) Land glaciation part in the earth's water balance, *IAHS/Unesco Symposium on World Water Balance, Reading*, IAHS Publication No. 92, Vol. 1, pp. 54–7.

Kukla, G., Hecht, A. and Wiesnet, D. (eds) (1981) *Snow Watch 1980, Glaciological Data*, World Data Center A for Glaciology (Snow and Ice), University of Colorado, Boulder, Colorado, USA, Report GD-11.

Kunzi, K.F., Patil, S. and Rott, H. (1982) Snow cover parameters retrieved from Nimbus-7 Scanning Multichannel Microwave Radiometer (SMMR) Data. *IEEE Trans. Geosci. Remote Sensing*, **GE-20** (4), 452–467.

Leaf, C.F. (1967) Areal extent of snow cover in relation to streamflow in Central Colorado, *International Hydrology Symposium, Fort Collins*, pp. 157–64.

Lenco, M. (1982) Télédétection et ressources naturelles (Remote sensing and natural resources). *Nature*, Unesco, Paris, **18** (2).

Linsley, R.K., Jr (1943) A simple procedure for the day-to-day forecasting of runoff from snowmelt. *Trans. Am. Geophys. Union*, Part III, pp. 62–7.

Martinec, J. (1973) Evaluation of air photos for snowmelt-runoff forecasts, *Unesco–WMO–IAHS Symposia on the Role of Snow and Ice in Hydrology*, Banff 1972, IAHS Publication No. 107, Vol. 2, pp. 915–26.

Martinec, J. (1977) Expected snow loads on structures from incomplete hydrological data. *J. Glaciol.* **19**, (81), 185–95.

Matson, J. and Wiesnet, D.R. (1981) New data base for climate studies. *Nature (London)*, **289** (5797), 451–6.

Matzler, C. and Schanda, E. (1983) *Snow Mapping with Active Microwave Sensors*, Int. Bericht, Institut fur angewandte Physik, Universitat Bern, Bern.

Meier, M.F. (1964) Ice and glaciers. In *Applied Hydrology*, (ed. V.T. Chow) Section 16, McGraw-Hill, New York.

Mellor, M. (1964) *Snow and Ice on the Earth's Surface*, Cold Regions Research and Engineering Laboratory, Hanover, New Hampshire, Report II-C1.

Peck, E.L., Bissel, V.C., Jones, E.B. and Burge, D.L. (1971) Evaluation of snow water equivalent by airborne measurement of passive terrestrial gamma radiation. *Water Resour. Res.*, **7** (5), 1151–9.

Potts, H.L. (1937) Snow surveys and runoff forecasting from photographs. *Trans. Am. Geophys. Union*, South Continental Divide Snow-Survey Conference, pp. 658–60.

Potts, H.L. (1944) A photographic snow-survey method of forecasting runoff. *Trans. Am. Geophys. Union*, Part 1, pp. 149–53.

Rango, A. (1980) Remote sensing of snow covered area for runoff modelling, *Hydrological Forecasting, Proceedings of the Oxford Symposium*, IAHS Publication No. 129, pp. 291–7.

Rango, A. and Itten, K.I. (1976) Satellite potentials in snowcover monitoring and runoff prediction. *Nord. Hydrol*, No. 7, pp. 209–30.

Rango, A., Martinec, J., Foster, J. and Marks, D. (1984) Resolution in operational remote sensing of snow cover, *Proceedings Symposium on Hydrological Applications of Remote Sensing and Remote Data Transmission*, IUGG–IAHS General Assembly, Hamburg, 1983.

Unesco/IAHS/WMO (1970) *Seasonal Snow Cover*, Technical papers in hydrology, No. 2, Paris.

US Army Corps of Engineers, North Pacific Division (1956) *Snow Hydrology*, Portland, Oregon, pp. 123–43.

Volker, A. (1970) Water in the world. Public lecture on the occasion of the IAHS Symposium on Representative and Experimental Basins, Wellington.

Wilson, W.T. (1941) An outline of the thermodynamics of snow-melt. *Trans. Am. Geophys. Union*, Part 1, pp. 182–95.

4

Applications of remotely derived snow data

4.1 Hydrological importance of snow

One of the most important tasks of hydrology is river-flow forecasting. This is basic information for hydraulic engineering projects, such as water power generation, irrigation, municipal water supply, flood control and the planning of water management generally.

If a multiple-purpose reservoir is to maintain a certain storage capacity to leave room for flood control and at the same time keep a minimum water volume stored in order to improve the low flows, its successful operation depends on short-term forecasts of inflow. Seasonal forecasts are necessary for power generation: if a reservoir is excessively emptied in the winter months of a year with little snow, the water level necessary for normal operation is not restored in the spring by meltwater, so that the production of electricity falls below the target or even breaks down. In order to avoid this situation, reservoirs often maintain a safe water storage in the winter even if it means importing electricity from other sources to cover the demand. Then, especially in a year with abundant snow cover, reservoirs are quickly filled by meltwater and when the inflow exceeds the capacity of turbines, water must be released by emergency outlets and over the crest, without being used.

In alpine countries, artificial reservoirs are frequently situated above 1000 m a.s.l. and the highest parts of the catchment areas reach 3000–4000 m a.s.l. In such altitudes, snow is a dominant runoff factor. It not only represents more than 50% of the annual precipitation, but also contributes to runoff by a higher proportion than rain due to lower losses by evaporation and sublimation as compared with losses occurring during rainfall runoff. It may sometimes be misleading to illustrate the hydrological importance of snow by its proportion in the annual precipitation. At the Weissfluhjoch station, the seasonal snow cover accumulates, as a 30-year average (1951–1980), 800 mm of water. This corresponds to 55% of the annual precipitation which amounts to 1460 mm after correction of the catch deficit. Since there is no snowmelt in the winter months, 800 mm practically represent precipitation from October through April. The remaining 660 mm consists of rainfalls as well as snowfalls which can form a temporary snow cover even in the summer months. Consequently, the proportion of all snow in the annual precipitation may well exceed 70% at this high elevation. However, it is only the seasonal snow cover with its 55% which facilitates the seasonal runoff forecasts. The low evaporation losses also refer to this snow cover and not to the summer snowfalls.

It is evident that snow is more convenient than rain for river-flow forecasting because water is retained in a basin for weeks or months before being released. This gives the forecaster more time to evaluate the situation, including the

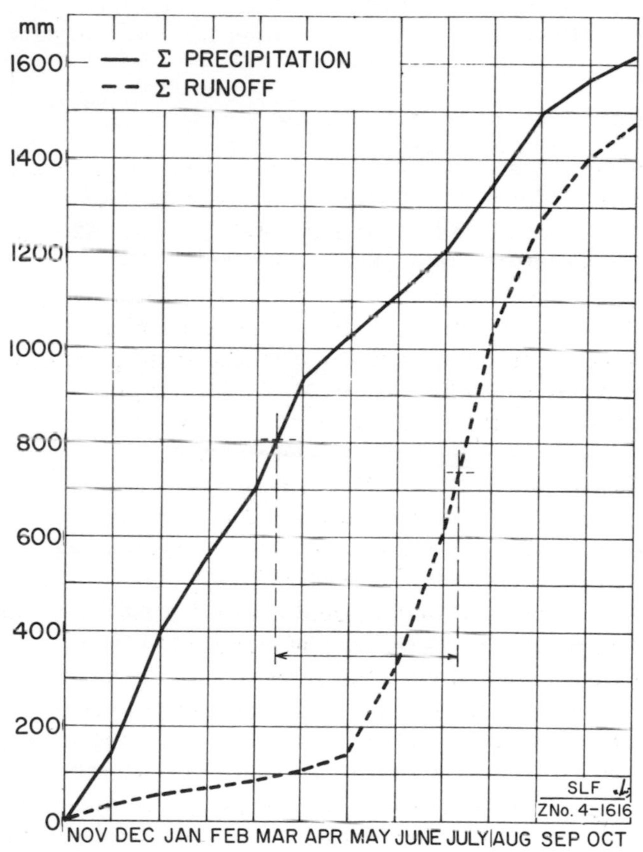

Fig. 4.1 Cumulative curves of precipitation and runoff in the alpine basin of Dischma, hydrological year 1967.

processing of remote sensing data, than for example in the event of a flash flood from a rainstorm. Figure 4.1 shows cumulative curves of precipitation with a major proportion of snow and runoff from the alpine basin of Dischma (43.3 km^2, 1668–3146 m a.s.l.) in the hydrological year 1967. While one half of the total precipitation amount is reached in March, the corresponding value for runoff follows 4 months later. Naturally, a certain part of this time lag must be attributed to the basin detention.

This seemingly comfortable time reserve does not mean that short-term changes and floods are excluded from the snowmelt season. It can always rain on top of snowmelt and the percolation of rainwater through the snowpack can be measured only in terms of minutes or hours. Some of the worst flood situations occur in such conditions or just after the disappearance of snow, when the soil is still soaked with meltwater.

Thus, the hydrological importance of snow can be seen not only in its quantitative presence in the hydrological cycle as outlined in Section 3.1, but also in its properties which enable remote sensing to be used for short-term and seasonal river-flow forecasts.

4.2 Snowmelt-runoff modelling

'Hydrologic models are mathematical formulations to simulate natural hydrological phenomena which are considered as processes or as systems' (Chow, 1964). A snowmelt-runoff model consists of two parts:

(1) A snowmelt model which calculates the snowmelt and rainfall over a basin.

(2) A transformation model which converts this input into the outflow from a basin.

As is explained in Section 3.4, the variable snow-covered area is essential for determining the areal snowmelt. Before the advent of remote sensing, snow-cover mapping was inefficient or impossible and so it is substituted in most models by an indirect approach: the snow cover is simulated from precipitation data, then melted by taking into account different temperatures and possibly other components of the energy input in the respective partial areas of a basin. The snow coverage results from the number and size of partial areas in which the simulated snowpack currently either still exists or has melted. In an international comparison of snowmelt-runoff models organized by the World Meteorological Organization in Geneva (WMO, 1982), nine models used this method and only one model was designed to use the remote sensing data on the real areal extent of the snow cover.

The performance of models was tested by comparing daily runoff values as computed by the models in six selected basins with the measured data. Characteristics of the test basins are listed in Table 4.1.

Table 4.1 Test basins for WMO's intercomparison of snowmelt-runoff models

Basin	*Country*	*Area (km^2)*	*Elevation range (m a.s.l.)*	*Test period hydrological years Oct.–Sept.*	*Data on areal extent of snow cover*
Durance	France	2170	786–4105	1970–79	Available in 1975–79 from snowline and Landsat
W-3	USA	8,7	346–695	1969–78	Approx. derived from snow courses
Dunajec	Poland	680	577–2301	1971–80	Approx. available in 1976 from snowline
Dischma	Switzerland	43.3	1668–3146	1970–79	Complete data from aircraft orthophotographs and Landsat
Illecillewaet	Canada	1155	1155–3107	1967–76	No data (Landsat data only after 1976)
Kultsjon	Sweden	1109	540–1580	1970–79	No data

Not all models were run on all basins because the required data were not included in one or another data sets, or because complicated models need more time to handle the data. The model which uses remote sensing (SRM) (Martinec *et al.*, 1983) is relatively simple:

$$Q_{n+1} = c_n[a_n(T_n + \Delta T_n)S_n + P_n]\,\frac{A \times 0.01}{86\,400}(1 - k_{n+1}) + Q_n k_{n+1} \qquad (4.1)$$

where Q is the average daily discharge in $m^3\ s^{-1}$; c is the runoff coefficient expressing the losses as a ratio (runoff/precipitation); a is the degree-day factor (cm °$C^{-1}\ d^{-1}$) indicating the snowmelt depth resulting from 1 degree-day; T is the number of degree-days (°C d); ΔT is the adjustment by temperature lapse rate necessary because of the altitude difference between the temperature station and the average hypsometric elevation of the basin or zone; S is the ratio of the snow-covered area to the total area; P is the precipitation contributing to runoff (cm) (a preselected threshold temperature, T_{CRIT}, determines whether this contribution is rainfall and immediate); A is the area of the basin or zone in m^2; conversion from cm $m^2\ d^{-1}$ to $m^3\ s^{-1}$ = 0.01/86 400; k is the recession coefficient indicating the decline of discharge in a period without snowmelt or rainfall: $k = Q_{m+1}/Q_m$ ($m, m+1$ are the sequence of days during a true recession flow period); and n is the sequence of days during the discharge computation period. Equation (4.1) is written for a time lag between the daily temperature cycle and the resulting discharge cycle of 18 hours. As a result, the number of degree-days measured on the nth day corresponds to the discharge on the $n + 1$ day. Different lag times will result in the proportioning of day n snowmelt between discharges occurring on days n, $n + 1$ and possibly $n + 2$.

The model can be adjusted to any wide elevation range of a basin in the following general form:

$$Q_{n+1} = (I_{An} + I_{Bn} + I_{Cn})(1 - k_{n+1})\,Q_n k_{n+1} \qquad (4.2)$$

where I is the input from snowmelt and precipitation as outlined in Equation (4.1), computed separately for elevation zones A, B, C or any number of zones which may be necessary.

The only obstacle to running this model was the lack of snow-cover data in some basins and years. An example of runoff simulation in the Durance River basin is shown in Fig. 4.2. Apart from visual inspection of the simulated hydrograph as compared with the measured runoff, the Nash–Sutcliffe coefficient (Nash and Sutcliffe, 1970) is one of the frequently used numerical criteria for evaluating the performance of a model:

$$R^2 = 1 - \frac{\sum_{i=1}^{n} (Q_i - Q'_i)^2}{\sum_{i=1}^{n} (Q_i - \bar{Q})^2} \qquad (4.3)$$

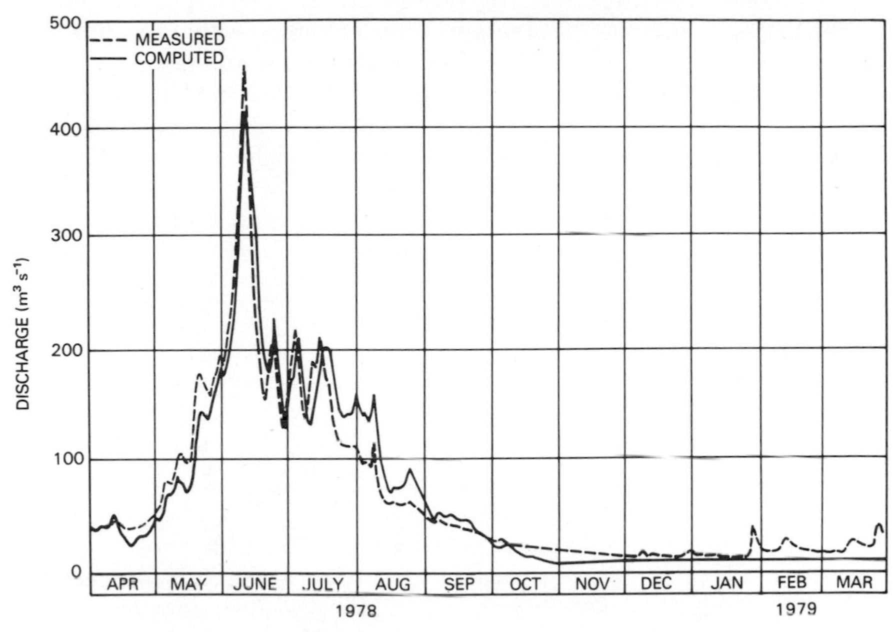

Fig. 4.2 Comparison between the simulated and measured daily flows in the Durance River basin, France, 1 April, 1978–31 March, 1979.

where R^2 is a measure of model efficiency, Q is the measured daily discharge, Q' is the simulated daily discharge, $\bar{Q}$ is the mean of measured discharge and n is the number of daily discharge values.

Thus, the value of R^2 depends on the daily differences between the simulated and measured hydrograph as seen in Fig. 4.2. Identical hydrographs result in $R^2 = 1$, low decimal values indicate a lack of agreement. Results for the SRM model are listed in Table 4.2. The model could be run only in years in which at least partial data on snow coverage were available: all 10 years in W-3 and Dischma, 5 years in Durance, 1 year in Dunajec. Thus, the results for Dunajec have less weight than the results for the other basins. No distinction is made between the calibration and verification period as designated in the WMO project, because SRM is not calibrated in any year.

Apart from an average performance, the worst experienced result is also of

Table 4.2 Coefficient R^2 for the SRM model. Average values and minimum values for snowmelt seasons of the available years

	Durance 1975–79	*W-3 1969–78*	*Dunajec 1976*	*Dischma 1970–79*
Average	0.850	0.799	0.759	0.836
Minimum	0.726	0.697	0.759	0.666

interest since it indicates the ability of a model to cope with difficult runoff seasons and to avoid serious failures. Table 4.2 lists these results.

The average values are arithmetic means of values for the respective snowmelt seasons. If an overall average of discharge for all the years considered were to be substituted into Equation (4.3), generally higher values of R^2 would be obtained.

Table 4.2 shows that if remote sensing of snow cover is available, a simple model like SRM can achieve good and reliable results.

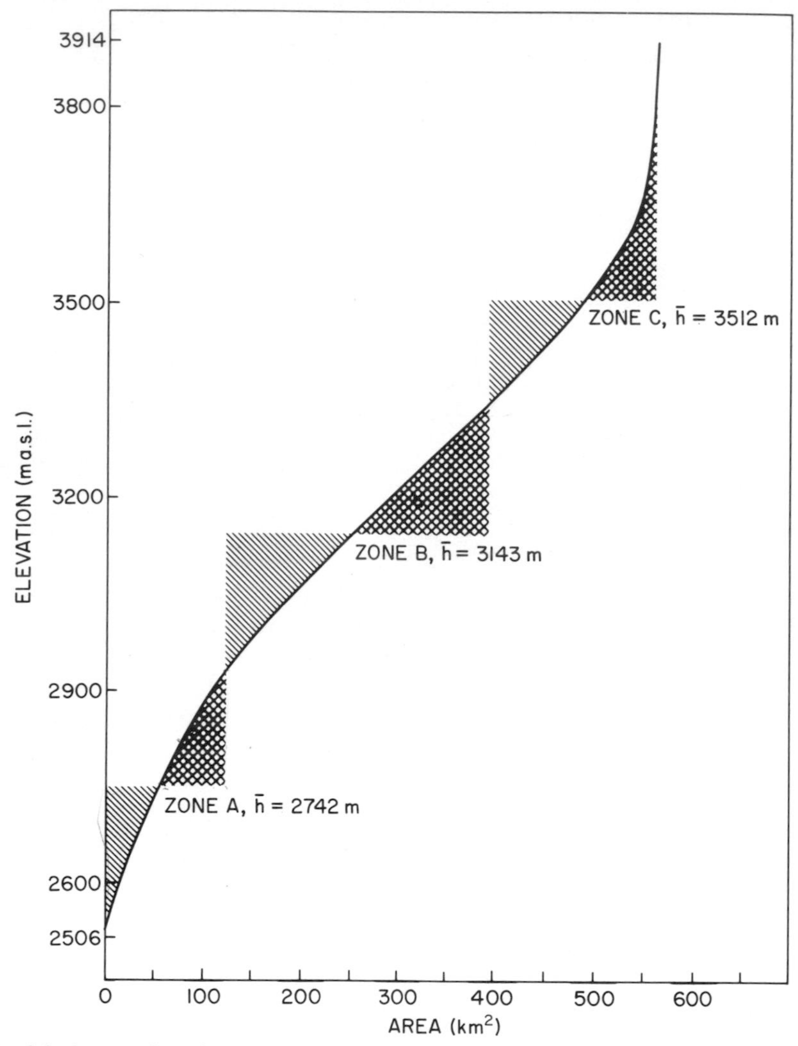

Fig. 4.3 Area—elevation curve and elevation zones in the basin of the river South Fork of the Rio Grande in Colorado, USA (Martinec *et al.*, 1983).

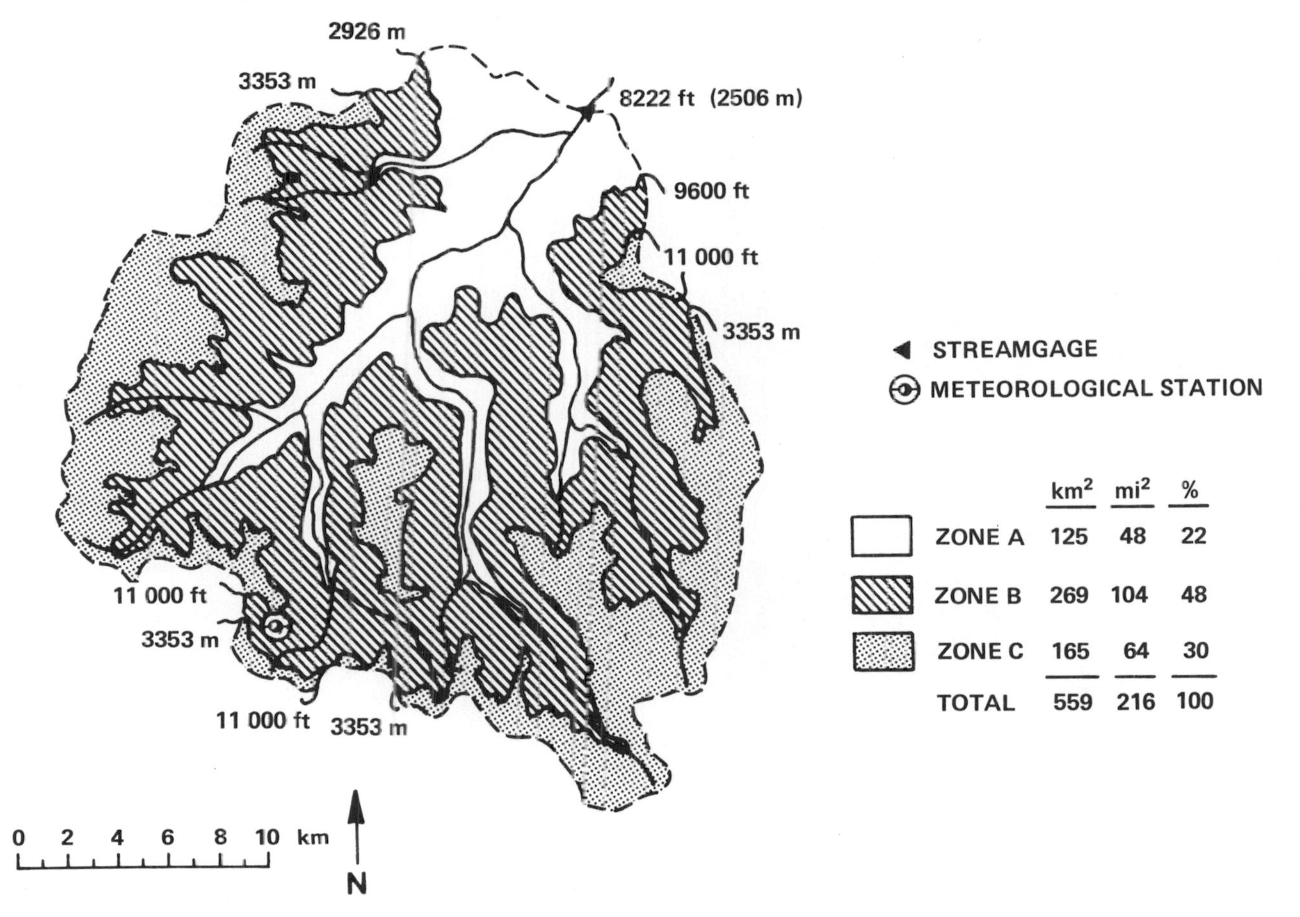

Fig. 4.4 Situation of the basin of the river South Fork of the Rio Grande with elevation zones (Martinec *et al.*, 1983).

The model simulation of the snow cover cannot replace, at least in the present stage, snow-cover measurements. Different snow covers are simulated by different models. A part of the winter precipitation is interpreted by one model as rain, by another model case as snow. Nevertheless, the models are able to simulate snowmelt runoff: to each simulated snow cover belong the corresponding model parameters optimized by discharge data.

The real snow cover signals by its persistence or quick disappearance extreme conditions in wet and dry years in which runoff deviates from the accustomed pattern. For example, by using Landsat data, streamflow in the South Fork of the Rio Grande in Colorado (559 km^2, 2506–3914 m a.s.l.) was simulated for two such difficult years by SRM (Shafer, 1980). Figure 4.3 shows the subdivision of the basin into three elevation zones and the determination of the mean hypsometric elevation for each zone. The degree-days needed in Equation (4.1) are extrapolated to these elevations by an appropriate lapse rate:

$$\Delta T = \gamma \frac{h_{st} - \bar{h}}{100} \tag{4.4}$$

where ΔT is the temperature adjustment (see Equation 4.1) (°C), γ is the temperature lapse rate in °C per 100 m, h_{st} is the altitude of the temperature station (m) and $\bar{h}$ is the zonal hypsometric altitude (m).

The areas of the elevation zones are shown in Fig. 4.4. The depletion curves of

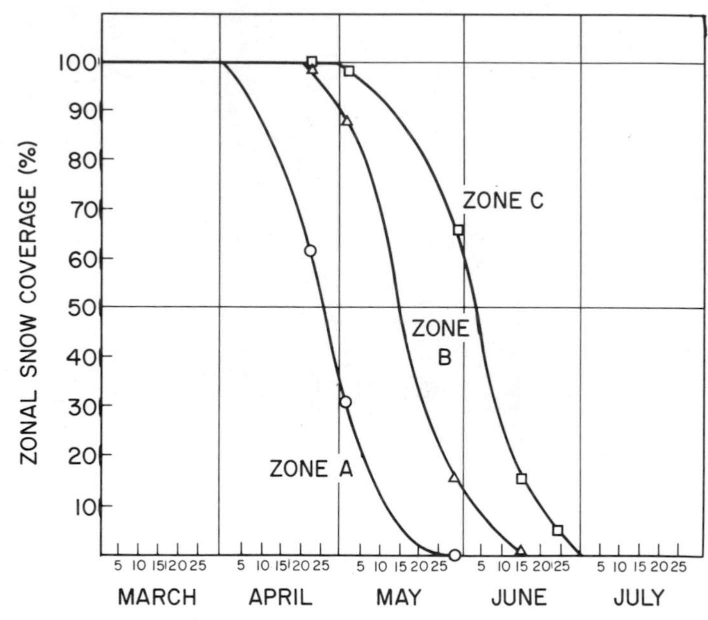

Fig. 4.5 Depletion curves of snow-covered areas in the basin of the river South Fork of the Rio Grande in 1976 (Martinec *et al.*, 1983).

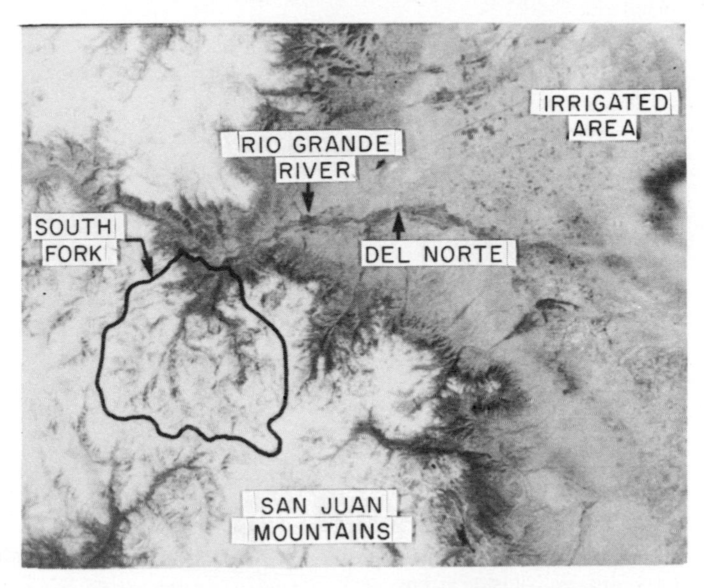

(a)

(b)

Fig. 4.6 Landsat images showing the snow cover in the catchment area of the Rio Grande River (Rango and Martinec, 1982): (a) 13 May, 1979; (b) 27 June, 1979.

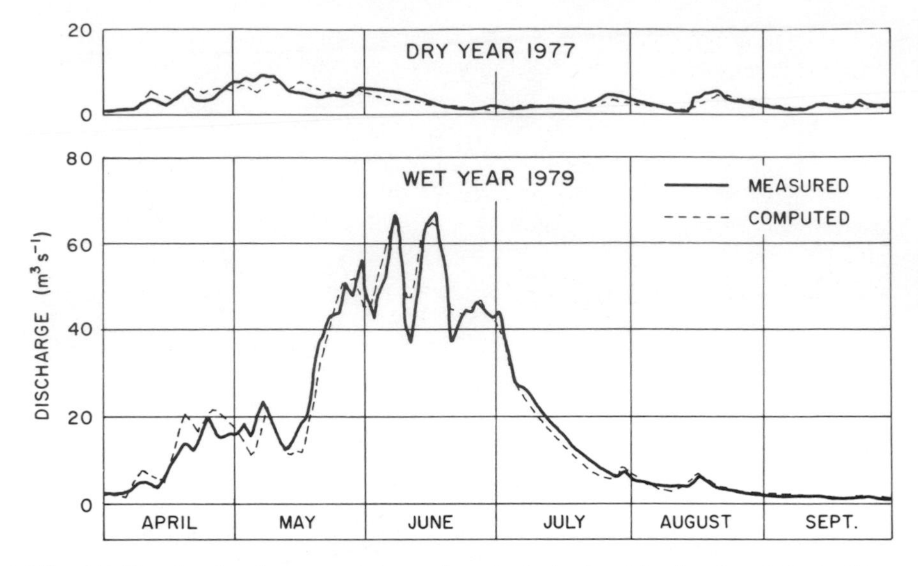

Fig. 4.7 Simulated and measured streamflows in the basin of the river South Fork of the Rio Grande: a comparison of two extreme years.

snow-covered areas have been derived from Landsat data. Figure 4.5 shows these curves as drawn in 1976 which was an average year. An example of Landsat images of the South Fork basin is shown in Fig. 4.6 (Rango and Martinec, 1982). With snow-covered areas read off the depletion curves, the temperatures and precipitation were the only other variables required to run the model.

Figure 4.7 shows the simulated and measured streamflows in two extreme years. In 1979, the runoff volume was six times greater than in 1977 and these volumes were also accurately simulated.

In line with efforts already in progress (Peck *et al.*, 1984), the use of remotely sensed snow-cover data should be encouraged in the interest of more reliable and simple snowmelt-runoff modelling.

4.3 Discharge forecasts

Snowmelt-runoff models are usually developed and tested in selected, especially well-equipped basins, in which hydrometeorological data are of good quality. Such favorable conditions are seldom encountered in basins where discharge forecasts are needed for practical purposes. Frequently, temperatures must be extrapolated from a distant station and precipitation data are not representative for the whole elevation range. Data for the more complicated models, such as the soil moisture, evaporation, incoming and outgoing radiation and wind speed, are hardly available at all.

For operational forecasts in real time, the pertinent data must be not only currently measured, but forecast as well. Therefore, simple models have a better chance to be transformed from the simulation mode to the forecasting mode. Since temperature forecasts are available, most models prefer the degree-day approach for calculating snowmelt, even if an energy balance subroutine is available. In the WMO project referred to in Section 4.3, nine of ten models used this method in one way or another, although the quality of data sets was above the average.

In the forecasting mode, the depletion curves of snow-covered areas cannot be used so conveniently as in the runoff simulations mentioned. Even if the satellite or aircraft data are promptly evaluated so that the current situation is known, the future course of a depletion curve must be extrapolated, then updated by the next available remotely sensed data.

The shape of a depletion curve in a given basin is determined by the initial accumulation of snow and by the changing melt-rates which depend mainly on the air temperature. Conventional depletion curves, which relate the percentage of a basin or zone covered by snow to the elapsed time during the snowmelt season, do not enable these effects to be identified. Therefore, modified curves have been introduced (Martinec, 1980), in which the time scale is replaced by accumulated degree-days. Since the effect of temperature fluctuations in the different weeks or months is eliminated, the decline of these curves is steep in a year with a shallow snow cover and slow in a year with a deep snow cover. Such extremes are illustrated in Fig. 4.8 for the elevation zone C of

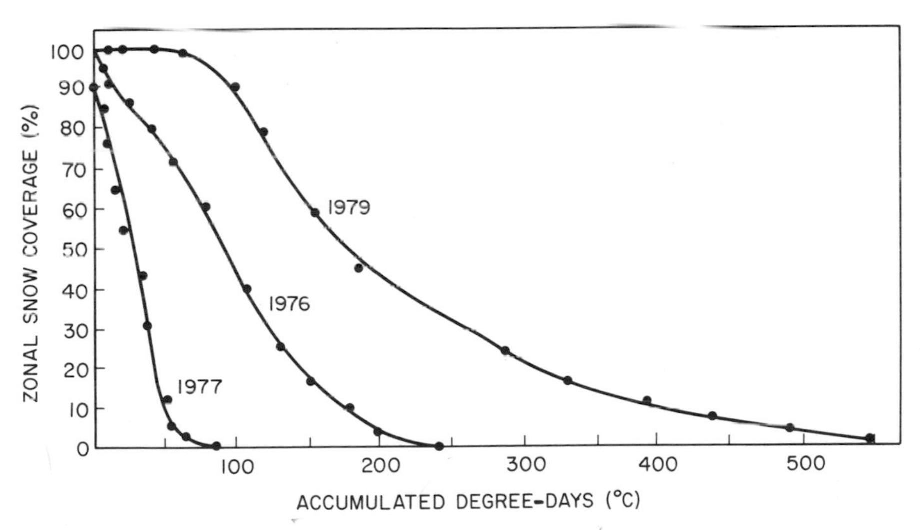

Fig. 4.8 Modified depletion curves of the snow coverage in the elevation zone C of the basin of the river South Fork of the Rio Grande: comparison of the decline in different years (Martinec *et al.*, 1983).

the South Fork basin in Colorado (Rango and Martinec, 1982). Recalling Fig. 4.7, the differing snow reserves in 1977 and 1979 as indicated by the modified depletion curves are confirmed by the resulting runoff volumes. If a Landsat image showing for example a 50% snow coverage were to be used to estimate the stored water volume, large errors would result: in 1979, this volume was evidently several times greater than in 1977, in accordance with the total degree-days needed to melt this snow, as well as with regard to the measured runoff volumes.

In some cases, these deviations appear to be less significant (Meier, 1973; Odegaard and Østrem, 1977). In the large Himalayan basins, it was even possible to derive a relationship between the average snow-covered area in April and the runoff in April through July (Rango *et al.*, 1977). By using the NOAA imagery, operational forecasts of seasonal runoff volumes are carried out. A relation derived for the Kabul River basin is shown in Fig. 4.9.

Coming back to the problem of predicting the course of depletion curves for short-term discharge forecasts, Fig. 4.10 gives an overview of factors affecting these curves. All hypothetical alternatives refer to a given uniform accumulation of snow in a basin, for example to an average water equivalent at

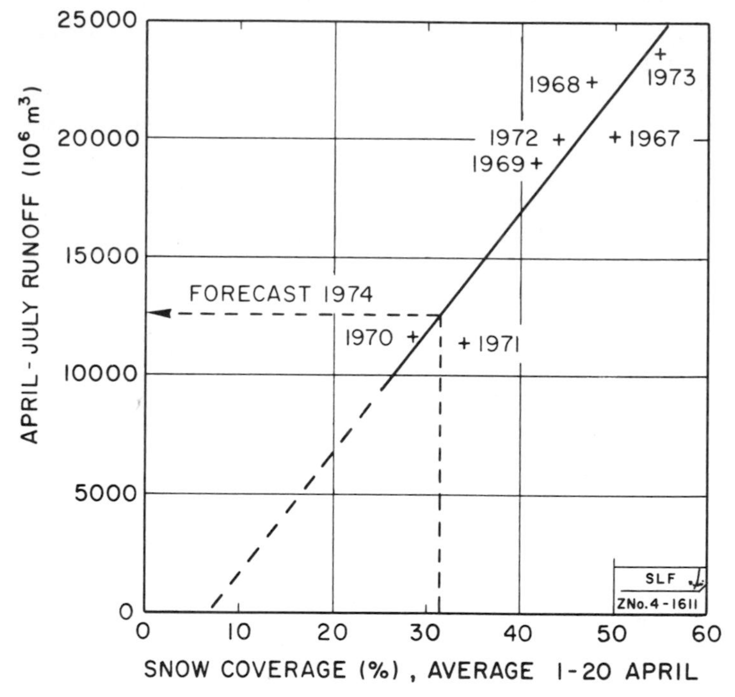

Fig. 4.9 Relation between the snow-covered area in April and the runoff volume in April, through July in the Kabul River basin, Pakistan, used for seasonal runoff forecasts. After Rango *et al.* (1977).

the start of the snowmelt season $H_{w\,max} = 60$ cm over an area of 100 km^2. Three types of weather which can occur during a snowmelt season in different years are considered:

(a) Average conditions: temperatures normal for the season, average number of snowfalls (new snow in the summer) which must melt before the ablation of the seasonal snow cover can continue. A delay of new snow represented by a step which is usually not noticed by a periodical monitoring of the snow cover.

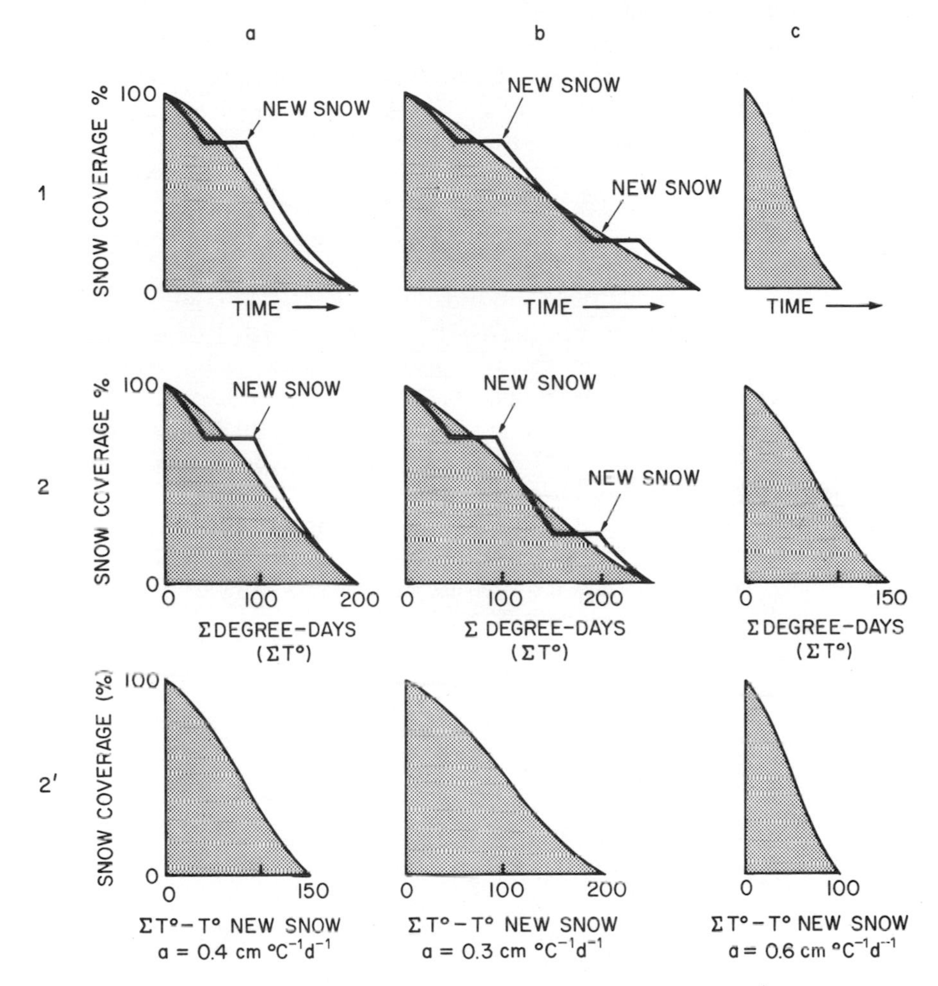

Fig. 4.10 Different types of depletion curves of snow-covered areas and their behavior in varying hydrometeorological conditions: 1, normal curves; 2, modified curves with time replaced by cumulative degree-days; 2′, degree-days used up to melt new snow subtracted from accumulated degree-days; a, average conditions; b, conditions delaying snowmelt; c, conditions accelerating snowmelt.

(b) Conditions slowing down the ablation of the seasonal snow cover; temperatures below normal, frequent snowfalls (steps).

(c) Conditions accelerating the ablation of the seasonal snow cover: temperatures above normal, no snowfalls during the snowmelt season.

As illustrated in Fig. 4.10, a conventional depletion curve (1) is greatly affected by these weather conditions. The same seasonal snow cover melts in a much shorter time in case 1c than 1b. The difference must be entirely attributed to the weather during the snowmelt season.

In the modified depletion curves mentioned (2), the time scale is replaced by accumulated degree-days. Therefore, the effect of temperature anomalies is eliminated. However, some differences between curves 2a, 2b and 2c remain due to the varying frequency of snowfalls (Fig. 4.10).

In the third version of depletion curves (2′), degree-days required to melt each new layer of snow during the snowmelt season are subtracted from the accumulated degree-days. The resulting relation between the number of degree-days applied to the seasonal snow cover and the snow coverage is independent of the weather types a, b and c and should be the same for a given accumulation of snow. However, degree-day ratios are variable, depending for example on the albedo, the thermal quality of snow, wind speed and intermittent rainfall. This effect is illustrated by assuming different degree-day ratios, which results in the curves 2′a, 2′b and 2′c (Fig. 4.10). The differences are exaggerated because the conditions which increase or reduce the degree-day ratios are not likely to persist during the whole snowmelt season.

In Fig. 4.11, the degree-days referring to the seasonal snow cover are replaced by accumulated snowmelt depths computed each day from the number of degree-days (with degree-days required for melting new snow subtracted) by

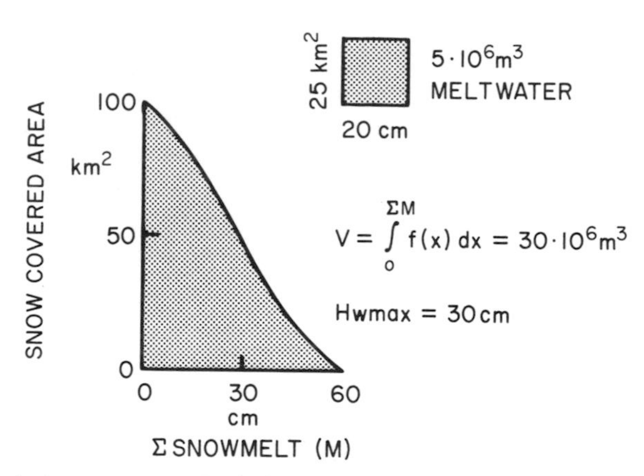

Fig. 4.11 Depletion curve type 3 relating snow-covered areas to accumulated snowmelt depths.

using the appropriate degree-day ratios. By relating each measured snow coverage to the total snowmelt depth from the start of the snowmelt season to the respective date of a satellite overflight, a depletion curve is obtained which is always the same for a given initial accumulation of snow in terms of the water equivalent. The integrated area between this curve and the *x*, *y* axis can be converted into the initial water volume stored in the snow cover, if the snow coverage is expressed in km^2 instead of as a percentage.

Figure 4.12 shows in more detail the separation of new snow from the seasonal snow cover. In runoff calculations, one part of the indicated snowfall is taken into account as the shaded area which delays the decline of the depletion curve. The other part of this snow which temporarily covered the snow-free area of the basin is taken into account as solid precipitation over this area, which is added to the computed snowmelt as soon as melting temperatures occur (Martinec *et al.*, 1983).

The 'snowmelt–snow coverage' depletion curve type 3 in Fig. 4.11 is more suitable for extrapolation by predicted temperatures than the curves type 2, 2′ in Fig. 4.10 because the distortions by different weather and snow conditions are reduced to a minimum. The procedure to be followed for operational discharge forecasts is illustrated by a hypothetical example illustrated in Fig. 4.13. The modified depletion curve (Fig. 4.11) always takes a different course according to the initial accumulation of snow. Two extreme curves for $H_{w\,max}$ = 20 cm and $H_{w\,max}$ = 60 cm represent an assumed range of conditions. Such curves must be predetermined for the given basin. In real time, if $H_{w\,max}$ on 1 April is not known and a snow coverage of 80% has been evaluated for example from a Landsat overflight on 15 May, the forecast for the next week is carried out as follows: (a) If the cumulative snowmelt to date is 30 cm, the curve for

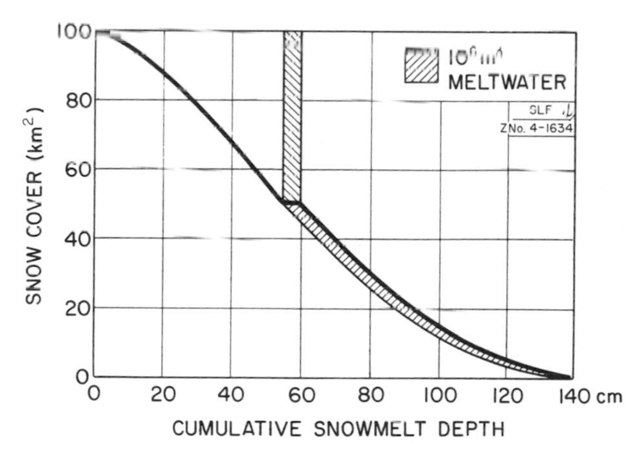

Fig. 4.12 The effect of new snow during the snowmelt season on the course of the depletion curve type 3.

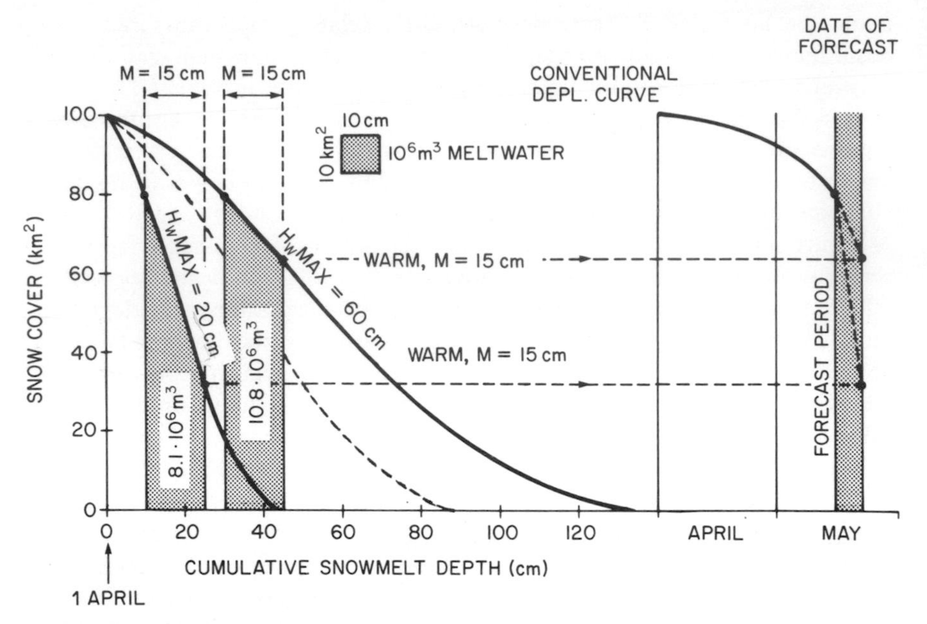

Fig. 4.13 Extrapolation of conventional depletion curves needed for real-time discharge forecasts using temperature forecasts and depletion curves type 3.

$H_{w\,max}$ = 60 cm is used. If a total snowmelt of 15 cm is calculated from the forecast temperatures and appropriate degree-day ratios for the coming week, the snow coverage will drop to 64% and this value is transferred to the conventional depletion curve on the right side. The daily values can be read off this extrapolated curve. (b) If the cumulative snowmelt to date is only 10 cm and still a snow coverage of 80% has been evaluated, this indicates low snow reserves and the curve for $H_{w\,max}$ = 20 cm is used. With the same predicted snowmelt of 15 cm, this curve will drop to 33% which is the extrapolated point of the conventional curve in this case. The daily values read off this curve are again used for day-to-day runoff computations.

Evidently it is not possible to use simply the snow coverage on 15 May and the predicted snowmelt depth. This would result in a meltwater volume of 12×10^6 m^3 while the forecast volume in the first case described is 10.8×10^6 m^3 and in the second case 8.1×10^6 m^3.

If $H_{w\,max}$ on 1 April is known, the modified curve to be used for extrapolation can be selected without waiting for the next satellite overflight. For points falling between the extreme curves, intermediate curves can be interpolated, each indicating a different initial accumulation of snow.

If temperatures below the freezing point are forecast, the predicted snowmelt $M = 0$ and the conventional depletion curve remains at 80% in all cases.

If it rains during the period of forecast, a corresponding amount must be added to the snowmelt depth as input to the runoff model. Unfortunately, quantitative rainfall forecasts hardly exist. Therefore, the expected rainfall depths are roughly estimated or, in the total absence of rainfall forecasts, statistically derived average values for the given season are substituted. Runoff computations are up-dated by actual rainfall amounts as soon as these are known.

If the temperatures originally forecast are changed during the week, the revised temperatures are used for a new extrapolation of the depletion curve which results in an up-dated runoff forecast.

The forecasting procedure illustrated by Fig. 4.13 is based on depletion curves type 3, which refer only to the original seasonal snow cover and not to additional snowfalls during the snowmelt season. This is necessary in order to determine the initial accumulation of snow for selecting the appropriate curve to be used in the given year for extrapolation. However, summer snowfalls may occur and delay the depletion of the seasonal snow cover as illustrated by Fig. 4.12: with the cumulative snowmelt depth of, for example, 80 cm, the snow coverage for the original seasonal snow cover would be 26%. Due to the effect of the new snow, it is 30%. It is this snow coverage which is seen by satellites and substituted into the snowmelt-runoff model. Figure 4.14 shows the transformation of snow-covered areas which disregard snowfalls during the snowmelt season (as is the case in Fig. 4.13) to actual snow-covered areas if such snowfalls occur. In this hypothetical example, a snowfall with a water equivalent $H_w = 10$ cm occurred in April, then melted but delayed the ablation of the seasonal snow cover as indicated by the dashed line on the left side. On the date of the forecast, the snowmelt depths add up to 35 cm (new snow excluded), which corresponds by the depletion curve type 3, valid for the given year, to a snow coverage of 60%. The actual snow coverage corresponds to the cumulative snowmelt depth of $35 - 10 = 25$ cm and is read off the dashed equidistant curve. For the forecast snowmelt depth of 15 cm, the snow coverage would drop, according to the depletion curve type 3, from 60% to 33% but due to the snowfall in April, the actual values are 74% and 50%. The snow-covered areas are transferred to the conventional depletion curve on the right and the extrapolated curve is used for day-to-day runoff computations. If the appropriate curve type 3 (belonging to the accumulation of snow in the given year) is not yet known on 1 May, the procedure is as follows: the conventional depletion curve indicates, according to measurements, $S = 74\%$ on 1 May. This value is plotted on the left side of Fig. 4.14 against the cumulative snowmelt depth (including new snow) to obtain a point on the dashed line. By plotting this value against the cumulative snowmelt depth excluding new snow (degree-days necessary to melt new snow are subtracted), a point of the curve type 3 is obtained and the curve valid for the given year is identified. The equidistant dashed curve can then be drawn and used for extrapolation of the conventional curve on the right by temperature forecasts.

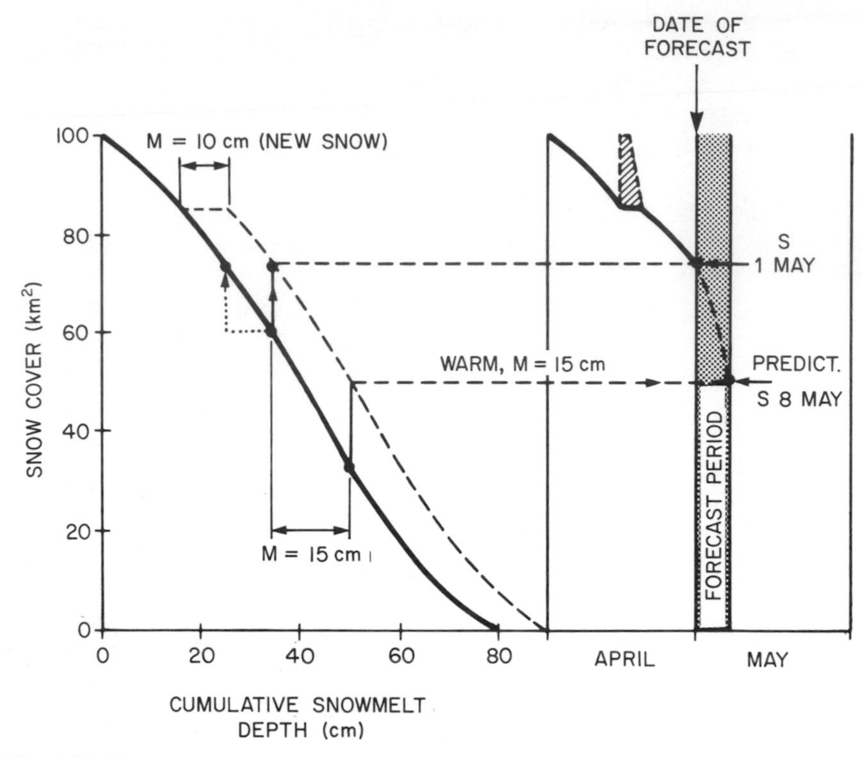

Fig. 4.14 Extrapolation of conventional depletion curves by temperature forecasts and depletion curves type 3, taking into account the effect of snowfalls during the snowmelt season.

The difference between the depletion curves type 2 and type 3 referred to in Figs 4.10, 4.11 is illustrated in more detail by Fig. 4.15. The pertinent calculations are summarized in Table 4.3.

For example, an average temperature of +1°C in April means 30 degree-days (°C d) and multiplied by the degree-day factor 0.3 (cm °C^{-1} d^{-1}) gives a monthly depth of meltwater of 9 cm. In Fig. 4.15, the calculated monthly volumes correspond to the shaded areas below the depletion curve type 3. For a conventional depletion curve with a normal time scale, all months would have equal lengths (if the difference between 30 and 31 days is neglected). For depletion curves type 2 with cumulative degree-days, July would be three times longer than April. For the curve type 3 with cumulative snowmelt depths, it is six times longer because the degree-day factor is double.

The outlined transformations of depletion curves of snow-covered areas may seem unnecessarily complicated to professionals specializing in remote sensing. But for an optimum hydrological exploitation of satellite data, the new space

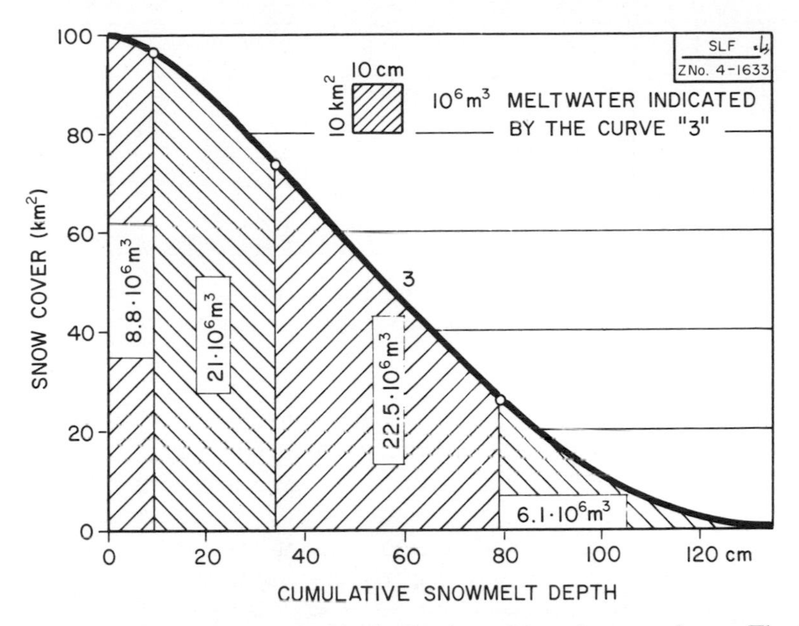

Fig. 4.15 Depletion curve type 3 and calculated monthly meltwater volumes. The time scale of conventional depletion curves appears condensed or stretched in the respective months.

Table 4.3 Computations of monthly meltwater volumes from a depletion curve of snow-covered areas

Month	*Average temperature (°C)*	*Average snow-covered area (km^2)*	*Degree-day factor ($cm\,°C^{-1}\,d^{-1}$)*	*Meltwater*	
				Depth (cm)	*Volume ($\times 10^6\,m^3$)*
April	+1	98	0.3	9	8.8
May	+2	87	0.4	24.8	21.6
June	+3	50	0.5	45	22.5
July	+3	11	0.6	55.8	6.1

technology and sophisticated processing of data should be combined with adequate hydrological application methods.

4.4 Economic benefits

Papers and reports on remote sensing of snow frequently refer to hydrology and promise economic benefits from a better use of water. Methods and examples outlined in the previous text may contribute to improve the credibility of such statements which are sometimes vague and not always realistic.

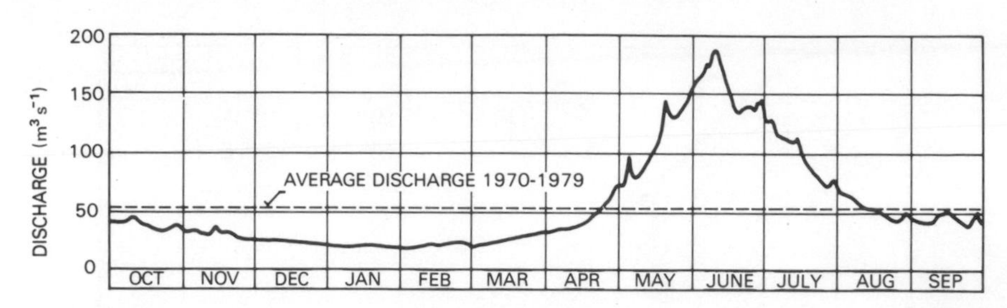

Fig. 4.16 Mean values of daily discharge in the Durance River basin, French Alps, from the period 1970–79 and long-term discharge average.

In order to prove an economic benefit, it is not sufficient to say that runoff forecasts are useful for electricity production, or that the damage from floods is great, or, from another field, that there are snow avalanche victims every year. For example, water power can be generated without runoff forecasts and such forecasts are even possible without remote sensing. The question is, how the present possibilities can be improved by remote sensing data.

The major users of water in the United States can be ranked by importance as follows (Castruccio *et al.*, 1980):

(1) Hydropower
(2) Irrigation
(3) Municipal and industrial water supply
(4) Navigation
(5) Recreation, land and wildlife management

The benefits for hydropower could be evaluated by comparing a reservoir operation based on non-existent or insufficient runoff forecasts with an improved operation and electricity production based on the best possible runoff forecasts.

Figure 4.16 shows runoff data from the Durance River basin (2170 km^2, 786–4105 m a.s.l.) which are available without snowmelt-runoff models:

(1) Long-term average of daily discharge
(2) Mean values of daily discharge for each day of the year.

The averaged runoff pattern sub 2 is sometimes called 'the Peasants Model'; it is thus a runoff simulation model which only needs a long period of discharge data to be developed for a given basin. It has been argued (Garrick *et al.*, 1978) that this simple approach can bring in some cases better results than complicated runoff models. Figure 4.17 shows, however, that deviations from the Peasants model in single years can be important and would certainly result in an inefficient hydropower production.

Satellite snow-cover data help to avoid forecasting failures in the years which

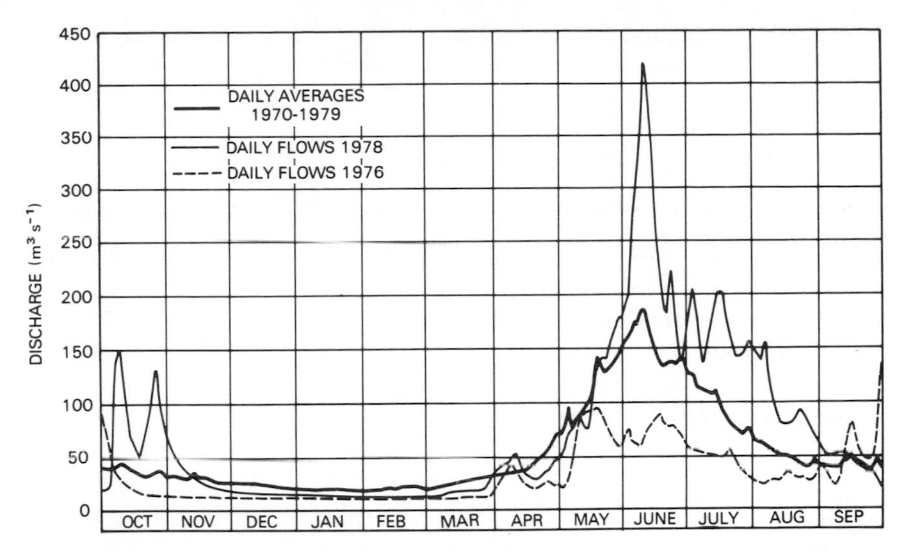

Fig. 4.17 Deviations from the 'Peasants model' for the Durance River basin in the years 1976 and 1978.

deviate from the usual runoff pattern, especially during the snowmelt season.

With regard to irrigation, improved discharge forecasts ensure a more efficient allocation of water in semi-arid climates which makes possible a higher production and an optimum crop selection.

The municipal and industrial water supply benefits by an improved efficiency of waterworks and by avoiding breakdowns in the water supply.

Navigation benefits can be expressed by the reduced transport costs if the use of waterways can be extended by improved scheduled releases of the reservoir water storage.

The benefits for recreation, land and wildlife management are mostly indirect and can be summarized as an improved quality of life which may affect the tourist industry.

Apart from benefits resulting from the value of water as a commodity, improved discharge forecasts serve for a more efficient flood control and thus reduce the harmful effects of water.

An extensive study was carried out (Castruccio *et al.*, 1980) with the aim of a quantitative assessment of benefits for hydropower and irrigation in the western United States. Basins in this area in which the snow cover was being mapped by satellites as of 1978 (Schneider, 1980) are shown in Fig. 4.18. The role of snow in water power generation is summarized for the 11 western states as follows (situation of 1978):

Annual water volume for hydropower	$2744 \times 10^9\,m^3$
Annual hydropower generation	186×10^9 kWh
Value of water for hydropower (3.8 cents/kWh)	US$ 7.15×10^9
Average proportion of snow	68%
Value of snow for hydropower	US$ 4.86×10^9

For irrigation, the following figures have been evaluated as totals for the 11 western states:

Annual water use for irrigation	$138 \times 10^9\,m^3$
Value of water (from crop value)	US$ 3.72×10^9
Average proportion of snow	71%
Proportion of surface water use	65%
Value of snow for irrigation	US$ 1.72×10^9

Fig. 4.18 Basins in the western United States with operational snow cover mapping by satellites as of 1978. Reproduced from Schneider (1980).

An estimated relative forecast improvement of 6% by the operational satellite measurement of snow-covered areas was used in computer benefit models (Castruccio *et al.*, 1980) in order to evaluate the benefit. Agriculture (irrigation) appears to be the major benefactor, receiving US \$ 28 × 10^6 annually. The benefit for hydroelectric production was evaluated at US \$ 10 × 10^6 per year. On the other side of a cost-benefit analysis, there are costs of satellite programs and of data processing. In making such comparisons, it should not be forgotten that satellites are also used for snow-cover mapping in other countries and are fulfilling many other tasks.

The relative importance of different areas of benefit varies in different parts of the world. For example, the role of water power in the total electricity production is more important in some countries than in the United States. In Canada, it amounted to 83% of the total production in 1962. In Switzerland, the hydroelectricity production in the early 80s fluctuates around 75% in spite of a considerable increase of atomic power and in Norway it represents nearly 100% of the total electricity production. It is a renewable source of energy and it can cover peak demands at short notice, in contrast to conventional or atomic thermal power plants.

In addition to the impact of real-time discharge forecasts, archived satellite data can be used to simulate runoff in ungaged areas. Such long-term discharge series are used for example to determine the optimum storage capacity of planned multiple-purpose reservoirs. Figure 4.19 shows a 5 year discharge series in the Durance River basin as simulated by the SRM model. The

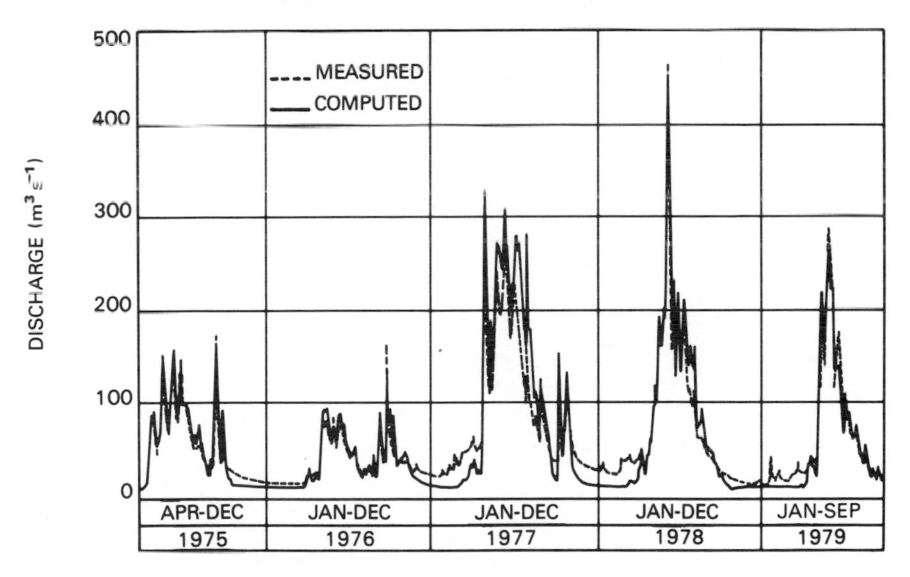

Fig. 4.19 River flows in the Durance River basin simulated from snow-cover data, precipitation and temperature for 5 years without up-dating by actual discharge.

measured values (which were available in this basin) are displayed to assess the accuracy of simulation. They have not been used for intermediate up-dating because this is not possible in an ungaged basin. The length of such simulated discharge series depends on the length of archived satellite data and on temperature and precipitation records which can be transferred and extrapolated from the next available station.

Apart from hydrological applications, large-scale snow-cover monitoring brings benefits with regard to harvest predictions and possible future changes of climate.

Although a complete quantitative evaluation is hardly possible, it can be concluded that a cost–benefit analysis supports continued and intensified efforts in the remote sensing of snow cover.

It is evident from the figures presented that the present achievements, although remarkable, are only a fraction of the potential benefits. Parallel development of hydrological models and procedures as well as improved interdisciplinary co-operation are the necessary conditions for obtaining the maximum exploitation.

References

Castruccio, P.A., Loats, H.L., Jr., Lloyd, D. and Newman, A.B. (1980) Cost/benefit analysis. In *Operational Applications of Satellite Snowcover Observations*, Workshop, Sparks, Nevada, 1979, NASA Conference Publication 2116, pp. 239–54.

Chow, V.T. (1964) Statistical and probability analysis of hydrologic data. In *Handbook of Applied Hydrology* (ed. V.T. Chow), McGraw-Hill, New York, Section 8, pp. 8–9.

Garrick, M., Cunnane, C. and Nash, J.E. (1978) A criterion of efficiency for rainfall-runoff models. *J. Hydrol*, **36**, 375–81.

Martinec, J. (1980) Snowmelt-runoff forecasts based on automatic temperature measurements, *Proceedings of the Oxford Symposium*, IAHS Publication No. 129, pp. 239–46.

Martinec, J., Rango, A. and Major, E. (1983) *The Snowmelt-Runoff Model (SRM) User's Manual*, NASA Reference Publication 1100, Scientific and Technical Information Branch.

Meier, M.F. (1973) Evaluation of ERTS imagery for mapping of changes of snowcover on land and on glaciers, *Symposium on Significant Results obtained from the Earth Resources Technology Satellite-1, New Carrolton, Maryland, NASA, Vol. 1, pp. 863–75.*

Nash, J.E. and Sutcliffe, J.V. (1970) River flow forecasting through conceptual models, Part 1 – A. Discussion of principles. *J. Hydrol.* **10**, (3), 282–90.

Ødegaard, H.A. and Østrem, G. (1977) *Application of Landsat Imagery for Snow Mapping in Norway*, Norwegian Water Resources and Electricity Board, Final Report, Landsat-2 Contract 29020, p. 20.

Peck, E.L., Johnson, E.R. and Keefer, T.N. (1984) Combining measurements of hydrological variables of various sampling geometrics and measurement accuracies, *Proceedings Symposium on Hydrological Applications of Remote Sensing and Remote*

Data Transmission, IUGG General Assembly, Hamburg, 1983, IAHS Publication (in press).

Rango, A. and Martinec, J. (1982) Snow accumulation derived from modified depletion curves of snow coverage, *Symposium on Hydrological Aspects of Alpine and High Mountain Areas in Exeter*, IAHS Publication No. 138, pp. 83–90.

Rango, A., Salomonson, V.V. and Foster, J.L. (1977) Seasonal streamflow estimation in the Himalayan region employing meteorological satellite snow cover observations. *Water Resour. Res.*, **13**, (1), 109–12.

Schneider, S.R. (1980) The NOAA/NESS program for operational snow cover mapping: Preparing for the 1980's. In *Operational Applications of Satellite Snowcover Observations*, Workshop, Sparks, Nevada, 1979, NASA Conference Publication 2116, pp. 21–39.

Shafer, B.A. (1980) *Report on the Martinec Model Project*, USDA. Soil Conservation Service (Denver, CO) submitted to NASA/Goddard Space Flight Center, Greenbelt, MD.

WMO (1982) WMO project for the intercomparison of conceptual models of snowmelt runoff, *Symposium on Hydrological Aspects of Alpine and High Mountain Areas in Exeter*, IAHS Publication No. 138, pp. 193–202.

5

Lake and river ice

5.1 The importance of lake and river ice

The formation of lake and river ice is inevitable in northern environments. In many areas it brings shipping and transportation on inland waterways to a standstill for several months every year. More precise knowledge of ice thickness and extent can be used to extend the shipping season or to warn of danger due to the presence of unexpected ice. In addition to the impact on man, the presence or absence of ice on lakes and rivers can have a major influence on the ecology of a region. The presence of ice can govern the viability of fish life in a lake or river. For example, in northern Alaska, some lakes freeze completely to their beds and are not suitable for a fish population whereas other lakes do not, allowing fish to survive below ice during the winter. In addition, ice that forms unexpectedly in lakes and rivers in a particular year can kill existing fish populations.

In a river, the amount of turbulence affects the formation and thickness of the ice cover. When velocity exceeds a critical value, river flow becomes turbulent. The Reynolds number, N_r, is commonly used to distinguish between laminar and turbulent flow (Morisawa, 1968):

$$N_r = \rho \frac{VR}{\mu} \tag{5.1}$$

where ρ is the density, V is mean velocity, R is hydraulic radius and μ is viscosity. If the Reynolds number is greater than about 600, then the water is considered turbulent (Ashton, 1979). At high velocities, common to most rivers, mixing due to turbulence will supercool the water over the entire depth and distribute small ice crystals, called frazil ice, throughout the depth of the river. In very fast-flowing rivers, or portions of a river, the production of frazil ice may be considerable since a solid ice cover cannot form.

In addition to the direct impact of ice on human activities, regional variations in ice formation and thickness are of scientific interest in climate studies as well. Factors such as temperature, amount, timing and duration of snow cover also affect the presence, thickness and duration of ice cover on lakes and rivers.

Flooding of rivers can be caused or exacerbated by ice. Two relatively common occurrences in northern areas that are often linked to flooding are ice jams and aufeis. Ice jams can form during the break-up period when river waters rise beneath the ice cover thus breaking the river ice into large floes. As these floes drift downstream, they may become grounded on gravel bars or meanders of the main river channel and form a blockage to the normal flow of water. Flooding ensues upstream from the ice jam and on both sides of the river as water is diverted from the ice jam. Rapid erosion and deposition of sediments may result; dredging may be necessary to restore a navigable channel. In some parts of Alaska, ice jams block the flow of rivers each spring. Aufeis or superimposed ice, forms in a river when ground or river water continues to flow into the channel after a river ice cover has formed. As the river ice cover thickens and the ground freezes, the ice cover begins to coalesce with the frozen ground or permafrost. Groundwater continues to flow upward under cryostatic pressure and spreads out on top of the river ice cover and extends onto the surrounding floodplain. During the spring when the river ice begins to melt upstream, flooding may occur at the site of the aufeis because the aufeis will take longer to melt than the river ice.

Aircraft and satellite-derived information on lake and river ice can be an aid to the shipping and transportation industries. Other important applications include assessment of lake ice thickness and duration, and determination of the availability of fresh water beneath lake ice for water supply studies.

For shipping and transportation, it would be desirable to employ microwave data (passive and/or active) for studies of ice movement and ice thickness determination because microwave data can be obtained day or night and irrespective of cloud conditions. Most of the microwave freshwater ice studies have been conducted from aircraft because the resolution from passive microwave satellite sensors is not adequate. At this writing there is no satellite-borne imaging radar to use for ice movement and ice classification. Visible and near-infrared satellite data have also proven very useful for freshwater ice studies despite the fact that cloud cover is a hindrance to obtaining data. NOAA data are superior to Landsat data in many ways for operational studies because data are acquired daily. Together, Landsat and NOAA data can be useful for studying ice build-up and break-up patterns in many lakes and rivers. Additionally, Landsat data have been used to locate fresh water sources by locating aufeis fields and, thus, springs in rivers (Harden *et al.*, 1977). Potential flood areas resulting from ice jams and aufeis can be studied remotely as well.

5.2 Freshwater ice thickness studies

In this section we will elucidate two methods for quantitatively determining freshwater ice thickness on lakes and rivers using remote sensing: radiometry and impulse radar sounding. Microwave radiometry offers the capability for wide area coverage over large lakes and rivers. For detailed studies of specific portions of lakes and rivers, impulse radar sounding is a more precise method. Impulse radar sounding is an especially useful engineering tool for planning construction (e.g. bridges) in ice-prone rivers.

The brightness temperature of freshwater ice is determined primarily by its physical temperature, surface characteristics, thickness and inclusions as well as instrument parameters such as frequency, look angle and polarization (see Chapter 2). A uniform slab of freshwater ice will emit microwave radiation in a quantity proportional to its thickness. In clear ice where there are few scatterers, the emissivity is predominantly affected by the ice/air interface, the ice/water interface, and the ice thickness. Ice thickness models utilizing microwave theory must take into account the contribution of each parameter to the overall emissivity. In lake ice there may be many bubbles that may be unevenly distributed throughout the ice. The bubbles lower the emissivity of the lake ice and contribute to nonuniform brightness temperatures across the lake. The numerous air/ice interfaces (bubbles) within the ice sheet, and the ice/water interface at the bottom of the ice sheet scatter the microwave radiation and lower the emissivity. The ice/water interface represents a dielectric discontinuity because the dielectric constant is ~ 80 for water and ~ 3.2 for ice at the frequencies commonly used for such studies. For smooth surfaces, the emissivity of freshwater ice depends upon the attenuation coefficient (a function of temperature) and the mean thickness of the ice (Swift *et al.*, 1980a).

Longer wavelength microwave radiation can be sensed from deeper within the ice than can shorter wavelength radiation. For freshwater lake ice studies, short wavelengths, e.g. 0.81 cm (37 GHz) to 1.4 cm (22.22 GHz) sense the snow overlying the ice while longer wavelengths, e.g. 6.0 cm (5.0 GHz), 21.0 cm (1.4 GHz) and longer sense the entire thickness of the ice and, in general, the ice/water interface. At the short wavelengths the overlying snow is contributing more to the observed T_B than is the ice because snow grains are large enough with respect to the radiation to cause scattering. However, snow grains are generally not large enough, compared to 6.0 or 21.0 cm radiation to cause significant scattering.

In 1971, a multifrequency radiometer was flown over Bear Lake, Utah, a freshwater lake in the United States (Schmugge *et al.*, 1974). At the time of the overflight the ice was 25 cm thick with 15 cm of snow on top. T_Bs from the five wavelengths shown in Fig. 5.1 demonstrate that the T_B decreased with increasing wavelength (or increasing penetration through the ice). A similar response was reported for Walden Reservoir in Colorado (Hall *et al.*, 1981).

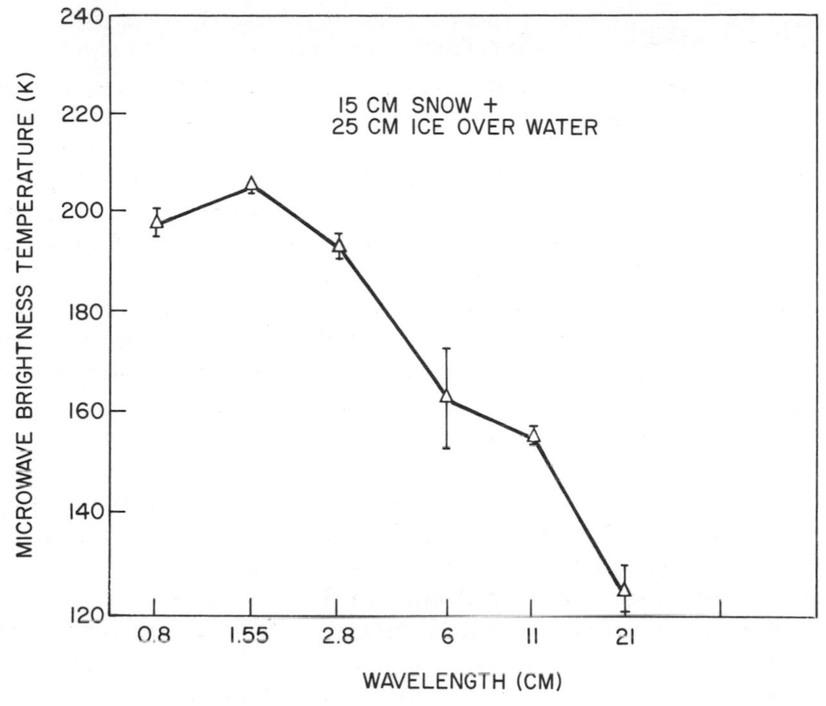

Fig. 5.1 Microwave responses to freshwater lake ice, Bear Lake, Utah (15 cm snow over 25 cm ice) – nadir viewing (from Schmugge *et al.*, 1974).

Figure 5.2 shows 6.0 cm vertically polarized radiometer data with a sharp (50 K) decrease in T_B between the snow-covered terrain to the 47 cm thick ice encountered on Walden Reservoir on 1 February, 1980. The snow depth overlying the lake ice was variable, ranging from 20 to 36 cm. For the Walden Reservoir study, the relationship between T_B and ice thickness using the 6.0 cm radiometer was found to be best using the nadir-viewing (0°) mode. The 6.0 cm horizontally and vertically polarized data obtained using 30° and 50° look angles showed poorer relationships with ice thickness than the nadir-viewed data (Table 5.1).

In the Mackinac Straits area of northern Michigan in the United States, a 6.0 cm radiometer was flown in a nadir-viewing mode on a NASA C-130 aircraft (Swift *et al.*, 1980a). In-situ ice thickness measurements were taken as well. T_Bs were generally found to range between 200 and 220 K for ice that was 60 to 80 cm thick. That range in T_B compares favorably with results of the Walden Reservoir study in which T_Bs for a 6.0 cm nadir-viewing radiometer also ranged between 200 and 220 K for ice that was between 65 and 70 cm thick. Figure 5.3 shows results of a flight over Lake Erie in March 1978 with microwave signatures collected at nine discrete center frequencies between 5522 and 5778

Table 5.1 Coefficient of determination (R^2) values for 6.0 cm sensor data over Walden Reservoir for different look angles and polarizations ($n = 4$) (from Hall *et al.*, 1981).

Angle	*Polarization*	R^2
0°	not applicable	0.98
30°	H	0.93
30°	V	0.94
50°	H	0.90
50°	V	0.91

MHz. Thick and thin ice are shown to be separated by a transition zone (Swift *et al.*, 1980b).

Clearly, passive microwave data can be expected to provide useful information on freshwater ice thickness and open water areas or leads. However, improved resolution relative to that currently obtainable from satellites is necessary for operational uses.

Impulse radar systems have been highly successful when used on the ground and in aircraft for quantitative measurement of ice thickness. As described in Chapter 2, an electromagnetic pulse is generated and reflections from the ice surface (or ice/air interface in the case of airborne measurements) and the ice/water interface are displayed in strip-chart format. Travel times of the electromagnetic radiation between interfaces are converted to ice thickness values. The travel times are based on a knowledge of the electrical properties of the ice and/or calibration of the instrument under conditions of varying ice thicknesses. The dielectric constant of lake ice, from 1 to 18 GHz, at −10°C in the Straits of Mackinac was found to be 3.17 for clear ice, 3.08 for milky ice and 2.99 for clear ice with large (> 0.6 cm) air bubbles (Cooper *et al.*, 1976). Knowing the dielectric constant, ϵ, of ice, the time delay, in nanoseconds can be

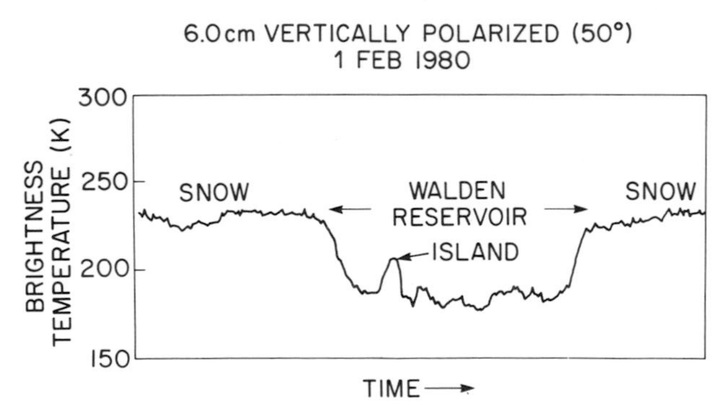

Fig. 5.2 T_B versus time on Walden Reservoir, Colorado (from Hall *et al.*, 1981).

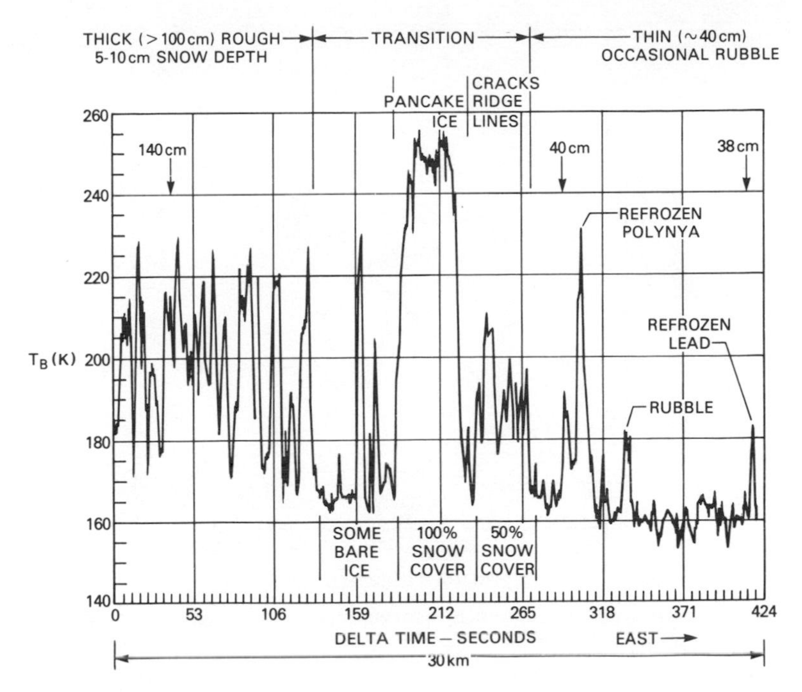

Fig. 5.3 T_B versus time for Mackinac Straits, Michigan (from Swift *et al.*, 1980).

directly related to ice thickness as seen in Equation (2.4) in Chapter 2.

On the Mackenzie River north of Inuvik, NWT in Canada, Campbell and Orange (1974) secured profiles of freshwater ice thickness by using an impulse radar on an all-terrain vehicle. Prior to this they had secured profiles of first-year and multiyear sea ice near the southern tip of Amund Ringnes Island in the Canadian Arctic. They found that the thickness of freshwater ice was easier to obtain than that of sea ice because the anisotropy of sea ice caused ambiguities in the data interpretation. Brine pockets and salt content of the sea ice are complicating factors not encountered in freshwater ice.

An impulse radar mounted on an all-terrain vehicle was used by Cooper *et al.* (1976) in the Great Lakes in March of 1975 to measure ice thickness. Twenty-five ice auger measurements were made in ice that varied from 20 to 60 cm in thickness. The ice auger measurements were found to compare very well with measurements obtained using the impulse radar (Fig. 5.4). The average error for the 25 measurements was 0.1%. Using aircraft data, Cooper *et al.* (1976) found that the aircraft system could not measure ice thickness less than 10 cm in thickness because the two return signals (1) from the ice/air interface, and (2) from the ice/water interface, could not be separated for such thin ice. Snow cover did not interfere with the accurate measurement of ice thickness at the

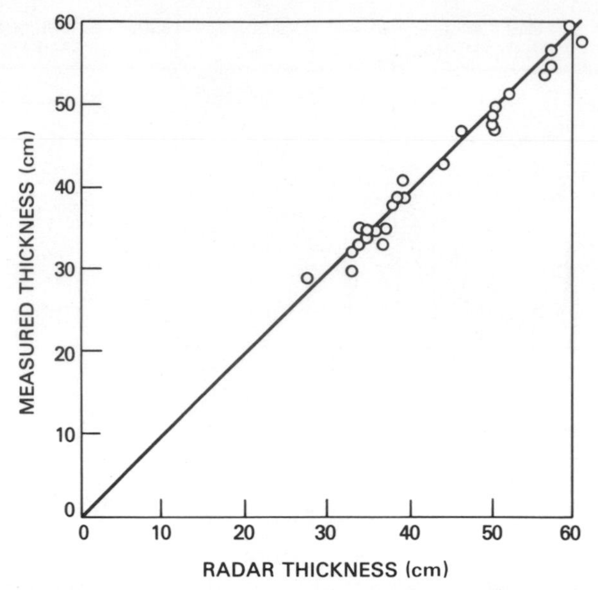

Fig. 5.4 Comparison of ice thickness measurements between auger and radar assuming $\epsilon = 3.1$ (from Cooper *et al.*, 1976).

frequencies used; however, slush or water on top of the ice, and rain prevented radar penetration.

The US Army Cold Regions Research and Engineering Laboratory (CRREL) has been employing a broadband impulse radar system since 1974 to secure profiles of ice from the ground and from the air. An antenna can be mounted on the side of a helicopter for use in remote sensing as shown in Fig. 5.5

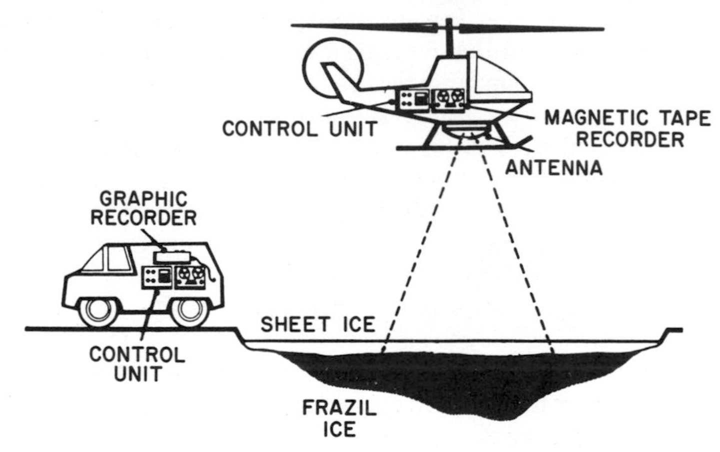

Fig. 5.5 Radar components during airborne collection of lake ice data (adapted from Dean, 1981).

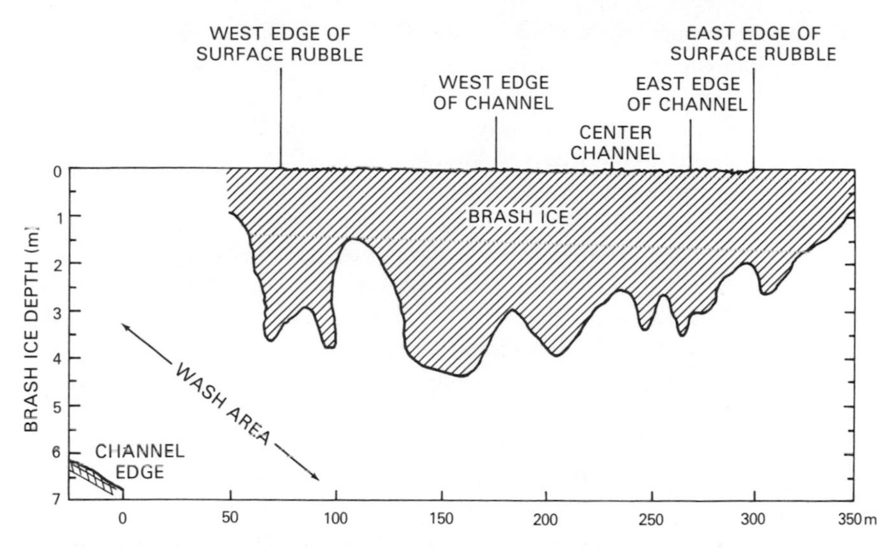

Fig. 5.6 Cross section of ice accumulation due to ship traffic on the St Mary's River near Sault Ste. Marie, Michigan – artist's sketch of interpreted short pulse radar data (from Dean, 1981).

(Dean, 1981). This system was able to secure a profile of an ice bridge across the Tanana River near Fairbanks, Alaska and to determine the bottom profile of a lake. In addition, ice accumulation on the St Mary's River near Sault Ste. Marie, Michigan, was clearly outlined as seen in Fig. 5.6.

Quantitative ice thickness measurements, such as those obtained using impulse radars, are useful for navigation and transportation in large lakes and rivers. It may also be of interest to determine whether a lake is frozen solidly to its bed. For example, during the winter of 1976, deep freezing caused a shortage of available fresh water for consumption in the Prudhoe Bay area in Alaska. Laborious drilling through 2 m thick lake ice was necessary to find unfrozen water sources (Kovaks, 1978). The operational use of impulse radar sounding to determine ice thickness could obviate the necessity for such laborious drilling. Further studies of lake depth/ice thickness determination are considered in the next section.

5.3 Lake depth and ice thickness studies in northern Alaska

The oriented lakes in northern Alaska dominate the landscape of the Arctic Coastal Plain. Tens of thousands of these shallow thaw lakes epitomize the very wet nature of the area. These lakes are formed from the thawing of ground ice under the influence both of geologic structure and the prevailing winds. Knowledge of the bathymetry of the lakes is important for water supply, wildlife and transportation considerations. Deep lakes that do not freeze to the bottom during the winter may be useful for winter water supplies and for stocking fish. Conversely, long, shallow lakes that do freeze to the bottom may be suitable as 'runways' for wide-body cargo planes that are sometimes required to transport supplies to the area (O'Lone, 1975).

Landsat has provided an excellent data source for studying many aspects of these lakes. Landsat data have been used to classify the lakes according to size, shape, degree of orientation, date of ice dissipation and stage of lake evolution. Additionally, use of Landsat and Seasat SAR data of these lakes can provide information relative to lake depth and ice thickness. Digital superposition of Seasat SAR imagery on MSS bands 5 and 7 imagery permits the lake evolution to be clearly illustrated. Marsh areas, former lake basins and emergent vegetation stand out clearly on the overlaid data as seen in Plate V (Hall and Ormsby, 1983). Imagery shown in Plate V was obtained during the summer of 1978 when Seasat SAR data were available for comparison with Landsat data.

Sellmann *et al.* (1975a) used 1973 Landsat imagery (band 7) to analyze sequential changes in ice cover on the oriented lakes and thus to infer lake depth on a regional scale on the Arctic Coastal Plain of Alaska. Ice cover on the deepest lakes takes longer to melt than ice cover on shallower lakes because the thicker ice on deep lakes requires more energy to melt it than does thinner ice. Three depth categories were determined: shallow (< 1 m deep), intermediate (1–2 m deep) and deep (> 2 m deep) lakes. Note in Fig. 5.7, MSS band 5 and 7 Landsat images acquired on 6 July, 1978 of the Teshekpuk Lake area, that a variety of spectral responses from the lakes is apparent. Some of the lakes are ice covered and others are ice free. Other signatures are intermediate in tone indicating that there may be some water overlying a floating ice cover. In addition, some lakes have a central ice cover surrounded by open water due to shallow benches on the peripheries of those lakes. Surficial water on the lake ice cover is more apparent on the band 7 image than on the band 5 image.

Side looking airborne radar (SLAR) imagery has been obtained over northern Alaska to observe the oriented lakes by several investigators (Sellmann *et al.*, 1975b; Elachi *et al.*, 1976; Weeks *et al.*, 1977, 1978 and 1981; Mellor, 1982). These investigators studied radar data at approximately the time of maximum ice thickness (March through early May) to determine relative ice thicknesses and lake depths. Two different categories of lakes rendered distinctly different radar responses: (1) lakes that were found to be frozen to the bottom gave low (dark) returns, and (2) lakes that were not frozen to the bottom gave high (bright) returns. This response has been observed using X- and L-band radars in northern Alaska. However, this effect was not found to be universal for all lakes.

The overriding explanation for the differential returns is that the contrast in dielectric constant between freshwater ice and water (ϵ = 3.2 and 80 respectively) is much higher than that between ice and frozen sediment (ϵ = 3.2

BAND 5

BAND 7

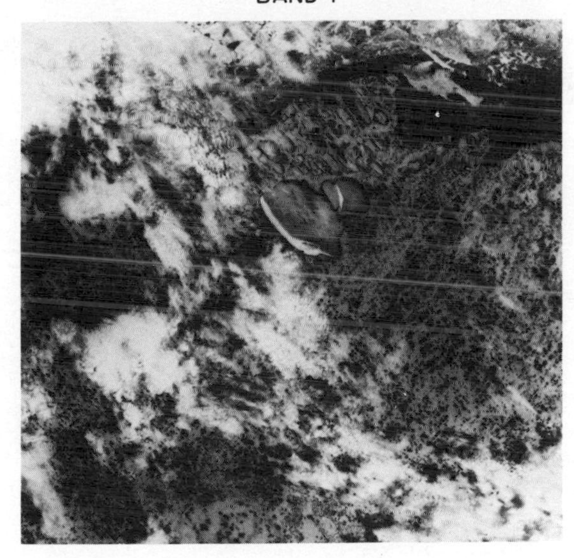

Fig. 5.7 Landsat-1 images of the Teshekpuk Lake region of the North Slope taken on 6 July, 1978 (I.D. 21261–21125). The band 5 image shows the extent of the ice cover on the lakes and the band 7 image shows lower reflectivities that are indicative of surficial melting.

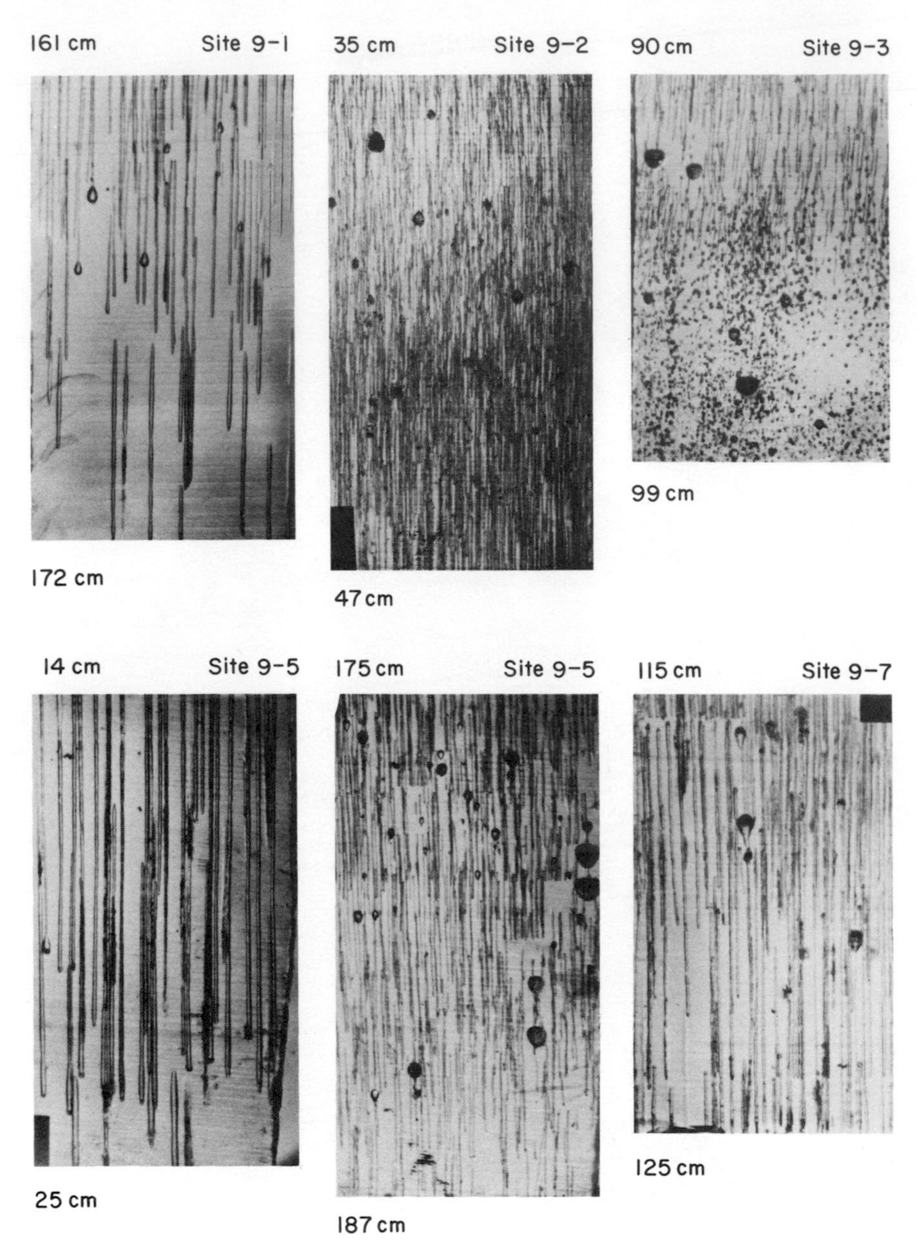

Fig. 5.8 Photographs (transmitted light) of 5 mm thick sections of lake ice from a number of sampling sites. Depth intervals are as indicated (from Weeks *et al.*, 1981).

to 8) (Elachi *et al.*, 1976). Stronger radar reflections occur at interfaces with a greater differential in dielectric constants. Thus lake ice that has water below will give higher radar returns than lake ice frozen to the lake bed. However, this alone cannot explain the effects seen on the North Slope lakes because at the non-perpendicular angles at which the radar beam encounters the ice, and in lieu of other scatterers, the radar beam would be reflected away from the sensor even with a highly reflective ice/water interface present. Weeks *et al.* (1978 and 1981) and Mellor (1982) have shown that the numerous elongated or columnar bubbles that are present in the ice on these lakes contribute to the high returns from the lakes that are not frozen to the bottom. Photographs of these bubbles are shown in Fig. 5.8 (Weeks *et al.*, 1981). (Locations of sampling sites are shown in Fig. 5.9.) The bubbles are oriented perpendicular to the forward scattering of the radar beam (toward the ice/water interface or ice/sediment interface). Bubbles with this orientation and adequate size allow enough

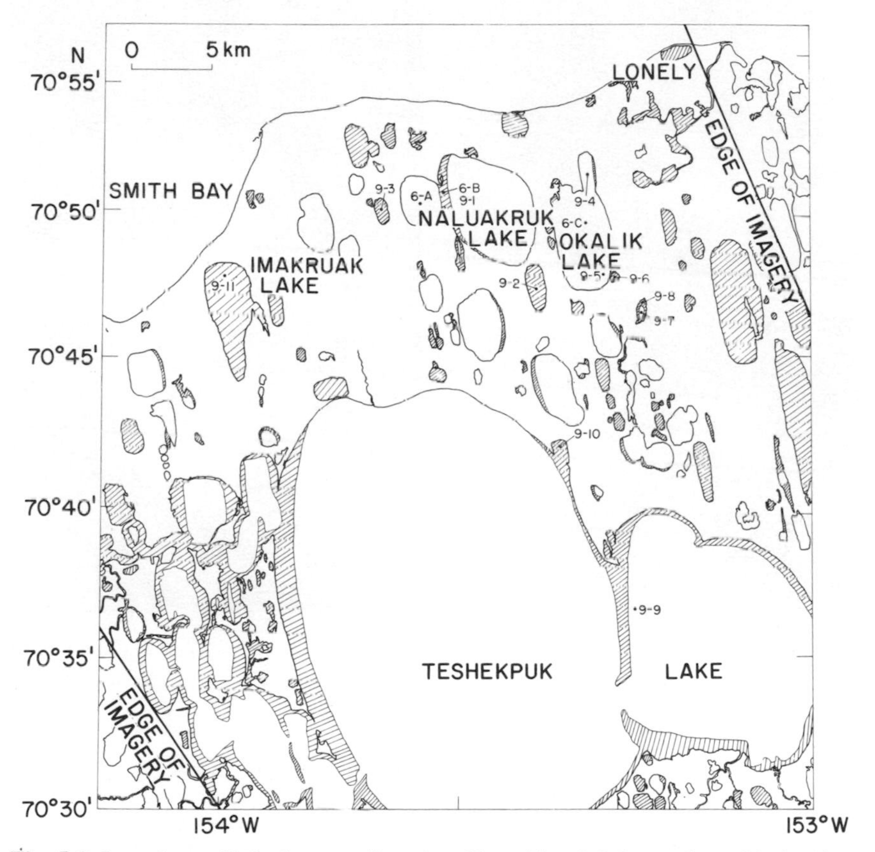

Fig. 5.9 Locations of lake ice sampling sites (from Fig. 5.8) in northern Alaska (from Weeks *et al.*, 1981).

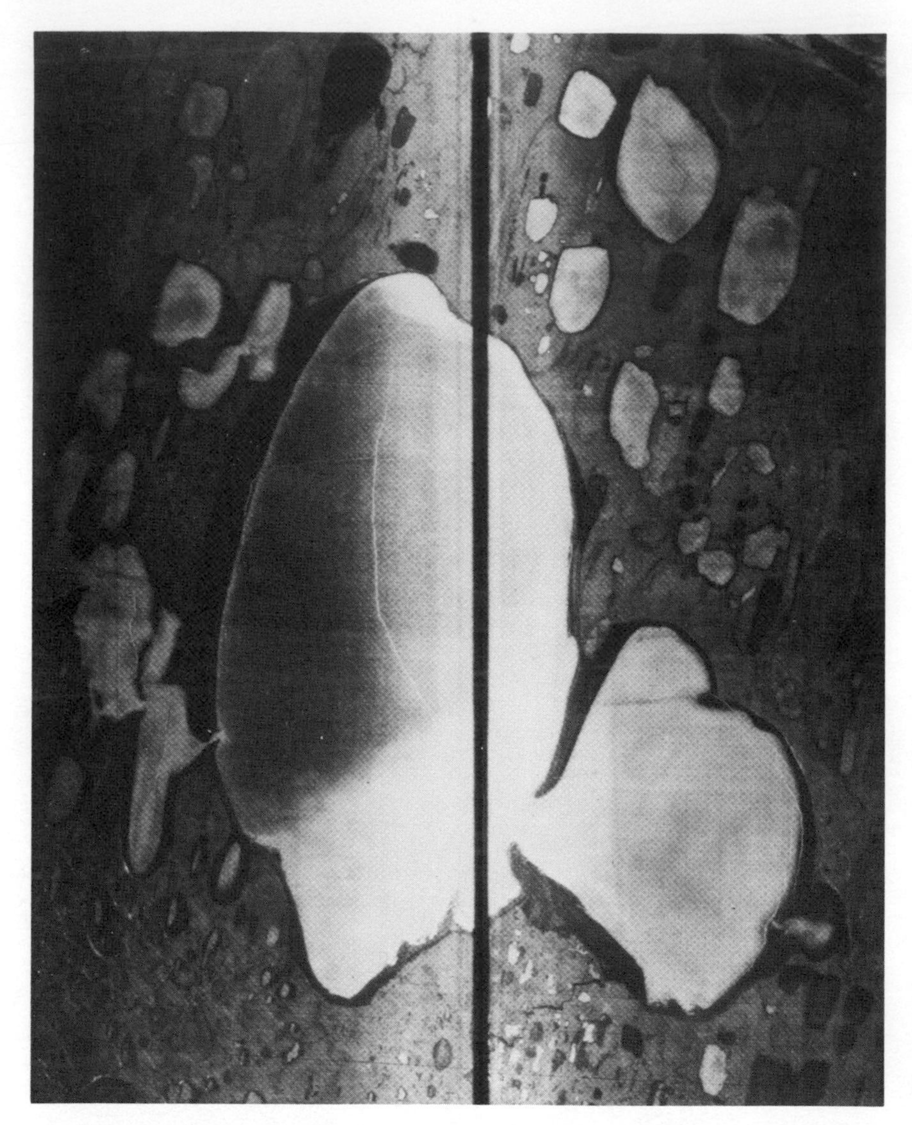

Fig. 5.10 SLAR imagery of the North Slope lakes near Lonely, Alaska. Bright areas indicate strong returns and dark areas indicate weak returns (from Weeks *et al.*, 1981).

additional scattering in lakes that are not frozen to their beds to cause the high X and L band radar returns (Weeks *et al.*, 1978 and 1981).

Interestingly, lakes or portions of lakes that are considered deep in northern Alaska (> 4 m) have few columnar bubbles as compared to the shallow lakes, and do not give high radar returns (even though they are not frozen to the bottom). Few bubbles will form when there is a substantial amount of water beneath the ice cover to accept the gases being emitted from the ice without the water becoming saturated (Mellor, 1982). The central and western portions of Teshekpuk Lake are considerably deeper than the rest of the lake, and though the ice is not frozen to the bottom, the radar reflections (X band in this case) are low (Fig. 5.10). Thus, to give the high radar returns, the lakes must (1) be generally shallow, (2) have numerous columnar air bubbles, and (3) not be frozen to the bottom. Most lakes on a world-wide basis do not fulfill all of these requirements and thus do not display the aforementioned radar returns that are characteristic of the North Slope lakes.

5.4 Ice in large lakes and estuaries

Visible, near-infrared and active microwave sensors provide important data for analysis of ice conditions in large lakes and estuaries. Landsat and SAR data can be used to classify ice types, and Landsat data can be helpful for inferring ice thickness. The role of wind and tidal currents in ice development and decay can be assessed using repetitive data of large areas. Such data, if disseminated in a near real-time manner, would be very useful to the shipping industry (Jirberg *et al.*, 1974; Bryan and Larson, 1975; Foster *et al.*, 1978; Foster, 1982).

Although presently operating aircraft SAR systems are not suited for quantitative ice thickness measurements, SAR imagery has been used to classify ice types. Bryan and Larson (1975) classified freshwater lake ice into several distinct categories using data from a multispectral radar operating at X-band and L-band in Whitefish Bay and the Straits of Mackinac, Michigan. The ice types identified ranged from smooth black ice which gave low radar returns to very rough ice with correspondingly higher (brighter) radar returns. In addition, rough ice areas were clearly visible on SAR imagery. Ridges containing rough ice gave very high returns, and though no quantitative assessment could be made of the ice thickness, knowledge of the location of such areas is useful to the shipping industry because rough ice is very difficult to negotiate (Jirberg *et al.*, 1974).

Landsat data can be used for inferring relative ice thickness by noting tonal differences in MSS imagery and reflectance changes in digital data. Landsat digital data were used to determine five ice thickness categories in the Chesapeake Bay on the east coast of the United States for the winter of 1976–77 (Foster *et al.*, 1978). During that particular winter, the Chesapeake Bay became 85% ice covered as compared to 10% in a normal year. This ice caused serious and unexpected problems for the local shipping and shellfish industries. Figures 5.11(a) and (b) show a Landsat image of the Chesapeake Bay and the MSS-derived ice thickness categories respectively. Distinct spectral differences were noted on the Landsat imagery and compared with in-situ ice thickness measurements. These spectral classes were found to correspond to the ice

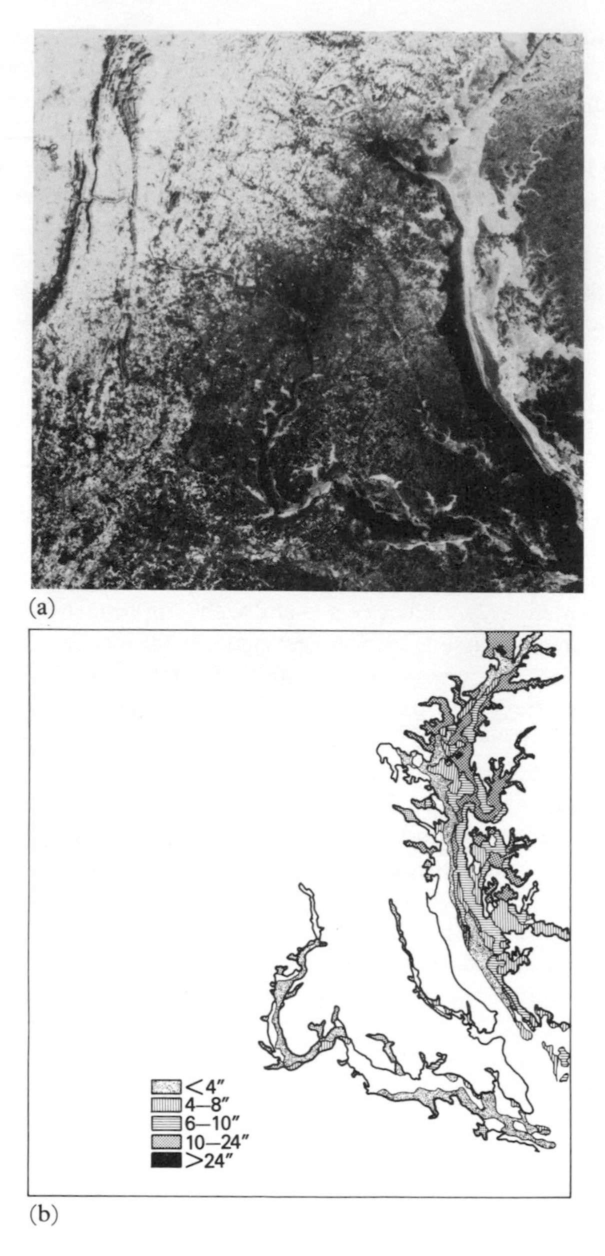

Fig. 5.11 (a) Landsat image of Chesapeake Bay ice – 2 February, 1977 (I.D. 2742–14534), (b) Map showing 2 February, 1977 Landsat-classified ice thickness data (from Foster *et al.*, 1978).

Table 5.2 Percentage of total ice type – Northern Green Bay and adjacent Lake Michigan waters as inferred from Landsat MSS digital data (after Leshkevich, 1981)

Ice category	*Percentage of total*	*km²*
Uncategorized	17.43	860.12
Snow-covered ice	20.84	1028.35
Open water	17.02	839.96
New (thin) ice	4.60	226.79
Consolidated pack	6.66	328.43
Slush	6.25	308.18
Brash	3.69	182.16
Flow	13.12	647.36
Wet snow-covered ice	10.40	513.03
Totals	100.00	4934.37

thickness measurements over relatively large areas in the bay (Foster *et al.*, 1978).

Also using Landsat digital data, Leshkevich (1981) was able to differentiate seven ice types on northern Green Bay and adjacent Lake Michigan waters using February 1975 data. New (thin) ice could be distinguished from water and ice and could be distinguished from thin cloud cover. The seven ice categories and the measurement of their total area coverage are shown in Table 5.2.

Landsat was also used to study ice patterns on the Chesapeake Bay for a 5-year period during the winters of 1977 through 1981. Table 5.3 shows the percentage of ice cover in the bay for each year as measured from Landsat imagery by Foster (1982). This particular 5-year period included some unusually cold winters. The Landsat imagery permitted detailed analysis of past ice patterns for the entire Chesapeake Bay. Ice conditions were analyzed in terms of the effects of meteorological conditions, i.e. temperature and winds, on ice formation, movement and decay. Thus, the overall pattern of ice development, movement and decay in the Chesapeake Bay is far better understood through analysis of the satellite imagery. This information can be used to assess future ice conditions

Table 5.3 Ice cover extent estimated from Landsat imagery and coast guard reports (from Foster, 1982)

	1977	*1978*	*1979*	*1980*	*1981*
Estimated percentage	85	30	60	15	50
Estimated date	10 Feb.	17 Feb.	20 Feb.	2 Mar.	18 Jan.

while the ice is in the process of forming. It may also reduce the adverse economic impact of unexpected ice formation on the transportation and shellfish industries.

The winter surface circulation and ice movement in Kachemak Bay, Alaska were analyzed using Landsat imagery for 1 November and 30 April for eight winters (1972–1980) (Gatto, 1981). A map of circulation patterns under varying wind conditions was also prepared. The impetus behind this work was the planned construction of a dam on the west end of the lake at the Bradley River and the fear that fresh water discharge into Kachemak Bay would increase causing additional ice to form in the bay, the impact of which is unknown. Glacial sediment, visible on MSS imagery, was used as a natural tracer to determine winter surface circulation. It was found that the circulation on inner Kachemak Bay is counterclockwise. Analysis of Landsat data enabled Gatto (1981) to conclude that if additional ice accumulates as a result of the dam, it will cause a greater concentration of ice along Homer Spit and the ice will be moved into the outer bay by northerly winter winds and surface currents. Again, the ability to anticipate increased ice concentrations in portions of the bay may help to aleviate possible adverse effects.

NOAA and Landsat data have been used a great deal for analysis of ice on the Great Lakes. Ice conditions are of great interest to the shipping industry and ice extent is important meteorologically because open water areas or large leads contribute moisture to feed major snow storms. Considerable analysis has been performed on the Great Lakes ice using NOAA satellite data (Wartha, 1977; Assel *et al.*, 1979; Wiesnet, 1979). Much of this work has dealt with analysis of ice conditions between years, comparison of ice conditions with meteorological conditions, and classification of lake ice. In addition, ice concentrations and thicknesses have been estimated from NOAA visible imagery.

Lake Erie is the shallowest of the Great Lakes and is subject to more extensive ice formation and earlier freezing than the other lakes. Wiesnet *et al.* (1974) used NOAA 2 VHRR infrared digital data to derive a thermal map of Lake Erie ice for 17 February, 1973. The temperatures were believed to be accurate within 2°C. Figure 5.12 shows results of this preliminary analysis of Lake Erie thermal patterns. The four 'transects' illustrate representative temperature distributions across the lake and they also revealed quite different ice surface temperatures. Transects A–A′ and B–B′ crossed over new, thin ice on the north side of the lake giving high surface temperatures (−2°C) relative to more southern portions of the lake where ice surface temperatures dropped below −8°C. D–D′ shows stable (−4°C) surface temperatures over the deep eastern part of the lake. The surface temperature of ice is affected by many factors such as air and water temperature, thermal conductivity of ice, and ice thickness and emissivity.

On Lake Huron, Strong (1974) found large surface water temperature variations that corresponded to different lake depths (from 4°C in the deep

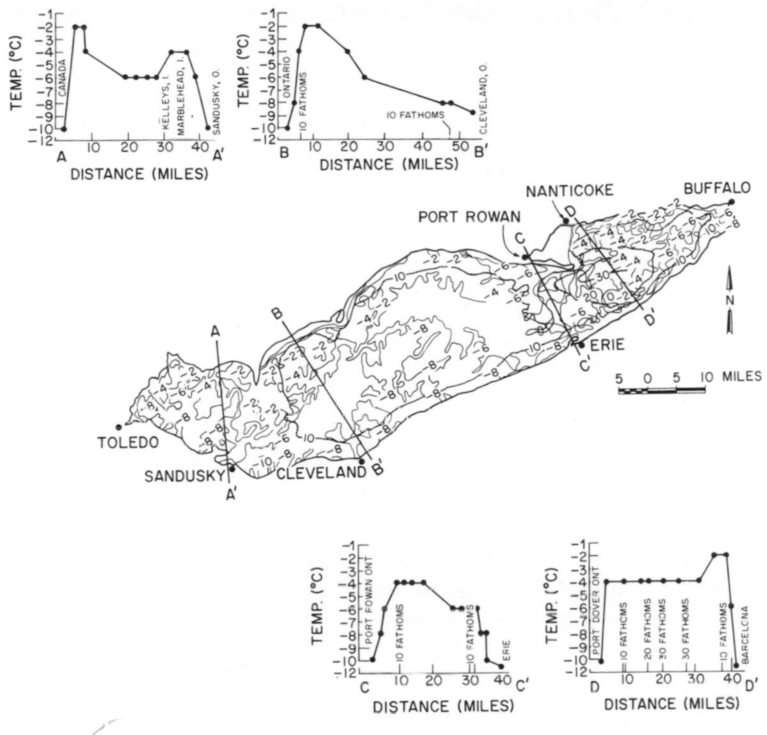

Fig. 5.12 Thermal map of the Lake Erie ice surfaces, 17 February, 1973, prepared from a digitized NOAA-2 VHRR-IR data format. Temperature variations across the lake are shown along four transects (from Wiesnet *et al.*, 1974).

northern basin to 20°C in the shallow Saginaw Bay) during the summer. Also observed was an area of relatively cold water upwelling on Lake Michigan that continued for about one week. This kind of information, obtained under ice-free conditions is useful for ice formation studies.

5.5 River ice break-up

River ice break-up can be a very dramatic occurrence in the melt season in many areas. Break-up can signify the imminent resumption of navigation in a river system but it can also cause extensive flooding and related problems. Since the advent of satellite imagery, Landsat and NOAA data have been employed to study the timing, nature and occurrence of river ice break-up as well as the flooding that is often associated with ice jams during the break-up period.

The ability to predict the date of river ice break-up would be an economic boon to many areas that rely on the river systems for transportation and

Table 5.4 Break-up dates along the Mackenzie River, 1975–77 (from Dey *et al.*, 1977)

	1975		1976		1977	
Station	*Satellite imagery*	*Field observation*	*Satellite imagery*	*Field observation*	*Satellite imagery*	*Field observation*
Fort Simpson	15 May	13 May	7 May	5 May	3 May	4 May
Wrigley	*	*	14 May	16 May	*	11 May
Norman Wells	*	27 May	15 May	12 May	15 May	15 May
Fort Good Hope	4 June	5 June	19 May	*	18 May	17 May
Arctic Red River	*	4 June	24 May	24 May	19 May	21 May
Napoiak	8 June	5 June	26 May	29 May	31 May	29 May

* Not available.

shipping. NOAA VHRR and AVHRR data are quite useful for monitoring river ice break-up especially when Landsat MSS data of selected areas and time periods are employed to augment the daily NOAA data. The NOAA satellites are advantageous for such studies because imagery of the same area is obtained daily – at a resolution of 1.1 km whereas each Landsat satellite obtains data from the same place on earth only once every 16 or 18 days. The Landsat frequency of coverage is not adequate for analysis of the progression of break-up. NOAA GOES data are of limited utility in Arctic and sub-Arctic areas because the inherent distortions present in the geostationary satellite imagery preclude the monitoring of rivers above 50° in latitude (McGinnis and Schneider, 1978).

NOAA and Landsat images of the Mackenzie River in the Northwest Territories in Canada were used to study the ice break-up patterns from 1975 to 1977. Dey *et al.* (1977) were able to detect the break-up pattern and record break-up dates through analysis of NOAA satellite imagery. And as can be seen from Table 5.4, their satellite-derived observations compared quite favorably with field observations.

NOAA 4 VHRR and GOES 1 visible imagery were used to monitor river ice break-up on the Ottawa River in 1976. Between 4 and 14 April, 1976, fourteen portions of the river were monitored daily using NOAA 4 and GOES 1 imagery. Eleven of the fourteen portions of the river were observed to break up in place and only three were observed to be ice covered at the end of the 10 day observation period (McGinnis and Schneider, 1978). Break-up of the tributaries was noted to be generally earlier than that of the main river channel. For this study, as with the Mackenzie River study, better resolution Landsat imagery (when available) was used for detailed analyses of portions of the river channel.

5.6 Ice jams and aufeis

As mentioned in Section 5.1, ice jams and aufeis can cause extensive flooding in northern areas. Remote sensing can be very useful in locating potential sites for ice jam and aufeis formation, and for measuring the extent of resultant flooding.

Several techniques have been employed either to alleviate (by blasting) or prevent (by dusting) the adverse effects of ice jams. The blasting technique is self explanatory and is performed after the ice jam has occurred. The dusting technique is a preventive measure in which sand or dust is spread on river ice and snow prior to break-up in order to lower the albedo of the ice and thus enhance ablation. Portions of the river channel are dusted in order to accelerate melting in specific locations prone to ice jams. Ice floes can then move freely through a channel opened by this technique (Miller and Osterkamp, 1978).

The effectiveness of the dusting technique is dependent upon precise pinpointing of the main river channel; Landsat imagery acquired in the fall can be useful for precisely locating the river channel. In Arctic areas, river channels migrate and may not be in the same location from year to year. Miller and Osterkamp (1978) used Landsat MSS band 4 imagery to precisely locate the main channel, and to locate probable sites for ice jams on the Yukon River (Fig. 5.13). Using enlargements of Landsat imagery, they were able to specify target areas for dusting. They concluded that Landsat data were potentially useful for ice jam prevention in Alaskan rivers and that conventional aerial photography was too expensive to use for routine monitoring of the thousands of miles of Alaskan rivers that are subject to ice jam formation.

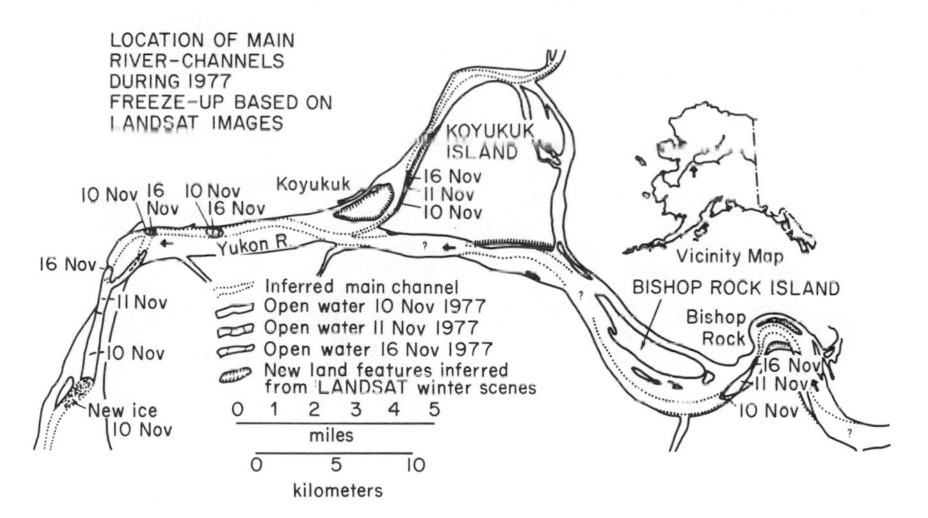

Fig. 5.13 River-channel locations determined from Landsat imagery for ice jam avoidance on the Yukon River, Alaska (adapted from Miller and Osterkamp, 1978).

In addition to flooding from ice jams, flooding during break-up can be caused by rivers overflowing sea ice, and by the presence of aufeis in rivers. Rivers of the North Slope of Alaska tend to melt out faster than the sea ice in the Arctic Ocean into which they flow. Analysis of Landsat imagery has shown that the North Slope rivers break up and overflow in sequence from southeast to northwest in response to variations in solar insolation at different latitudes (Barnes and Reimnitz, 1976). Using Landsat imagery, Barnes and Reimnitz (1976) were able to monitor and measure the development of flooding of the coastal sea ice at the mouth of the Sagavanirktok River in northern Alaska.

River aufeis, described in Section 5.1, can be both beneficial and damaging to human activity in the Arctic and sub-Arctic. It is beneficial in the sense that it is formed by spring water and thus its presence can be indicative of locations of springs which may be used for human freshwater consumption. Known springs in northern Alaska and Canada have been mapped and compared with locations of large aufeis fields as derived from Landsat imagery (Harden *et al.*, 1977; Hall

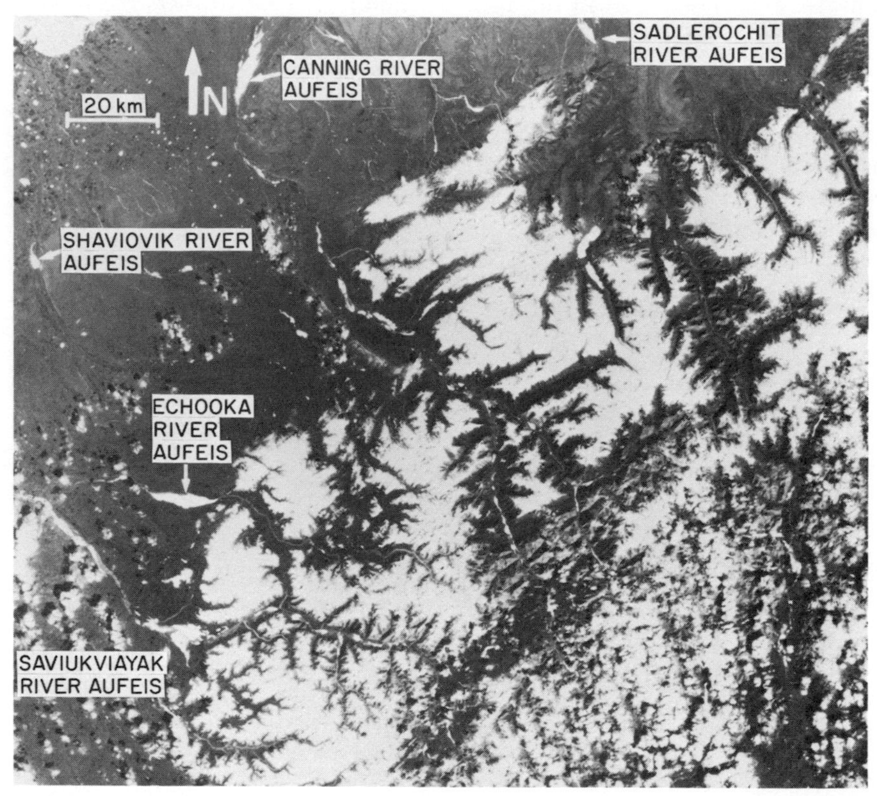

Fig. 5.14 Landsat band 5 image obtained on 15 June, 1979, showing aufeis on the North Slope of Alaska (I.D. 30467–20542).

and Roswell, 1981). Aufeis can also be very damaging to structures and towns due to flooding when aufeis melts or causes diversion of river waters onto the surrounding floodplain.

Landsat imagery is useful for locating large aufeis fields on a regional scale, for measuring its extent between years, and for monitoring its dissipation (Harden *et al.*, 1977; Hall and Roswell, 1981). Before break-up, Landsat MSS bands 6 and 7 (near-infrared) are most useful for locating aufeis in the snow- and ice-covered terrain because the absorption coefficient for ice is higher in the near-infrared bands (bands 6 and 7) than in the visible bands (bands 4 and 5). Thus, aufeis appears darker than the surrounding snow-covered terrain. After snowmelt, the visible bands display the best contrast between the highly reflective aufeis and the dark, often wet, surrounding tundra (Fig. 5.14).

Band 5 Landsat imagery was used to locate, measure and analyze inter-annual variations in the extent of aufeis by Hall (1980) for the years 1973 through 1979 for seven aufeis fields in the eastern Arctic Coastal Plain of Alaska. The aufeis under study was found to vary considerably among the years studied (Table 5.5). Figure 5.15 exemplifies this variability in the Canning River aufeis field in northern Alaska; the difference in maximum extent was 21.66 km^2 in the years studied. In June of 1974 the maximum aufeis extent has 4.4 km^2 and in June of 1979 the maximum aufeis extent was 26.0 km^2 as measured from Landsat MSS band 5 imagery. The variability of aufeis extent is dependent upon factors such as air temperature, amount of snowfall and groundwater supply. In some years, colder winter temperatures and/or light snowfall allow deeper freezing of the river ice and thus less room for the unrestricted flow of groundwater in the river

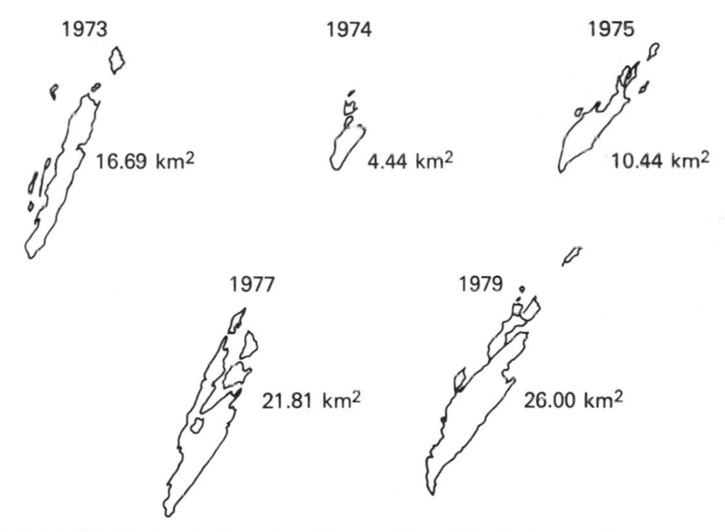

Fig. 5.15 Variability in the Canning River aufeis 1973–79 from available June Landsat data (adapted from Hall, 1980).

Table 5.5 Mean (km^2), standard deviation and coefficient of variation for June aufeis extent of seven Arctic Slope rivers (1973–79) (from Hall, 1980)

	Canning	*Echooka*	*Hulahula*	*Kongakut*	*Sadlerochit*	*Saviukviayak*	*Shaviovik*
Mean	15.88	22.23	3.10	23.85	6.69	15.91	2.96
Standard deviation	8.69	3.33	0.65	5.41	1.39	2.42	1.21
Coefficient of variation	54%	15%	21%	23%	21%	15%	41%

channel. Because of the inherent variability of aufeis, one cannot use one year of information to determine aufeis extent when considering construction in areas prone to aufeis development. Additionally, aufeis will not necessarily form in the same place in a river channel each year. Landsat data provide an historical record of aufeis extent variations back to July of 1972 and can be used for planning construction in areas prone to aufeis formation.

In addition to its obvious impact on engineering structures in the Arctic and sub-Arctic, aufeis is interesting geologically. Waters that form large aufeis fields in Siberia and Alaska emanate from deep groundwater, the location of which appears to be controlled by tectonic fracture patterns (Osokin, 1978; Hall and Roswell, 1981). The melted aufeis water has a quite different chemical composition from that of the adjacent river water (Hall and Roswell, 1981). Patches of pure calcium carbonate ($CaCO_3$) slush found on the Shaviovik and Canning River aufeis fields on the Arctic Coastal Plain are a further indication that the aufeis-forming water emanates from deep within the ground in calcareous bedrock that exists in the Brooks Range and foothills to the south of the Arctic Coastal Plain (Fig. 5.16). These $CaCO_3$ patches were found during field work in the summer of 1978 (Figs. 5.16, 5.17 and 5.18).

Flooding can result if aufeis fills a stream channel causing the streamflow to be diverted onto the floodplain. Flood waters will freeze in the winter, but will inundate surrounding areas with water from streamflow in the spring and summer months. Flooding resulting from aufeis is a major cause of highway damage, often resulting in washouts, and should be considered in the design of construction projects in many aufeis-prone regions.

Fig. 5.16 Calcium carbonate slush on the Shaviovik River aufeis field, 15 July, 1978 (Photograph by D. Hall).

Fig. 5.17 A sod-covered mass of aufeis on the Canning River that remained frozen as the surrounding ice ablated because of insulation afforded by the thick sod (photograph by I. Virsnieks).

Fig. 5.18 A ground photo of ablating aufeis in the Canning River in northern Alaska in July 1978 during field work (photograph by I. Virsnieks).

References

Ashton, G.D., (1979) River Ice. *Am. Sci.*, **67**, 38–45.

Assel, R.A., Boyce, D.E., DeWitt, B.H., Wartha, J. and Keyes, F.A. (1979)—*Summary of Great Lakes Weather and Ice Conditions Winter 1977–78*. National Oceanographic and Atmospheric Administration Technical Memorandum, ERL GLERL-26.

Barnes, P.W. and E. Reimnitz, (1976) Flooding of sea ice by the rivers of Northern Alaska. *ERTS-1 A New Window on Our Planet*, (eds R.S. Williams and W.D. Carter), U.S. Geological Survey Professional Paper 929, pp. 356–59.

Bryan, M.L. and R.W. Larson, (1975) The study of fresh-water lake ice using multiplexing imaging radar. *J. Glaciol*, **14**, 445–457.

Campbell, K.J. and A.S. Orange, (1974) Continuous sea and fresh water ice thickness profiling using an impulse radar system. *Advanced Concepts and Techniques in the Study of Snow and Ice Resources*, (eds H.S. Santeford and J.L. Smith), U.S.

International Hydrological Decade, National Academy of Sciences, Washington, DC, pp. 432–43.

Cooper, D.W., R.A. Mueller and R.J. Schertler, (1976) Remote profiling of lake ice using an S-band short-pulse radar aboard an all-terrain vehicle. *Radio Sci.*, **11**, 375–81.

Dean, A.M., Jr. (1981) *Electromagnetic Subsurface Measurements.* US Army Cold Regions Research and Engineering Laboratory, Hanover, NH, CRREL Special Report SR 81–23.

Dey, B., Moore, H. and Gregory, A.F. (1977) The use of satellite imagery for monitoring ice break-up along the Mackenzie River, NWT. *Arctic*, **30**, 234–42.

Elachi, C., Bryan, M.L. and Weeks, W.F. (1976) Imaging radar observations of frozen Arctic lakes. *Remote Sensing Environ*, **5**, 169–75.

Foster, J.L. (1982) Ice observations on the Chesapeake Bay 1977–1981. *Mariners Weather Log*, **26**, 66–71.

Foster, J.L., Schultz D. and Dallam, W.C. (1978) Ice conditions on the Chesapeake Bay as observed from Landsat during the winter of 1977, *Proceedings of the 1978 Eastern Snow Conference, 2–3 February, Hanover, NH, pp. 89–104.*

Gatto, L.W. (1981) Ice distribution and winter surface circulation patterns, Kachemak Bay, Alaska, *Proceedings of the International Geoscience and Remote Sensing Symposium (IGARSS'81), 8–10 June, 1981, Washington, DC*, (*IEEE Dig.*, **2** (1981) pp. 995–1001).

Hall, D.K. (1980) *Analysis of the Origin of Water which Forms large Aufeis Fields on the Arctic Slope of Alaska using Ground and Landsat Data*, PhD Dissertation, University of Maryland, College Park, MD.

Hall, D.K. and Roswell, C., (1981) The origin of water feeding icings on the eastern North Slope of Alaska. *Polar Rec.* **20**, 433–8.

Hall, D.K., Foster, J.L. Chang, A.T.C. and Rango, A. (1981) Freshwater ice thickness observations using passive microwave sensors. *IEEE Trans. Geosci. Remote Sensing*, **GE–19**, 189–93.

Hall, D.K. and Ormsby, J.P. (1983) Use of Seasat synthetic aperture radar and Landsat Multispectral Scanner Subsystem data for Alaskan glaciology studies. *J. Geophys. Res.*, **88**, 1597–607.

Harden D., Barnes, P. and Reimnitz, E. (1977) Distribution and character of naleds in northeastern Alaska. *Arctic*, **30**, 28–40.

Jirberg, R.J., Schertler, R.J., Gedney, R.T. and Mark, H. (1974) Application of SLAR for monitoring Great Lakes total ice cover. In *Advanced Concepts in the Study of Snow and Ice Resources* (eds H.S. Santeford and J.L. Smith), US International Hydrological Decade, National Academy of Sciences, Washington, DC, pp. 402–11.

Kovaks, A. (1978) Remote detection of water under ice-covered lakes on the North Slope of Alaska. *Arctic*, **31**, 448–58.

Leshkevich, G.A. (1981) *Categorization of Northern Green Bay Ice Cover using Landsat-1 Digital Data – a Case Study*, National Oceanographic and Atmospheric Administration, NOAA Technical Memorandum, ERL GLERL-33.

McGinnis, D.F. Jr., and Schneider, S.R. (1978) Monitoring river ice break-up from space. *Photogramm. Eng. Remote Sensing*, **44**, 57–68.

Mellor, J.C. (1982) *Bathymetry of Alaskan Arctic Lakes: a Key to Resource Inventory with Remote-sensing Methods*, PhD Dissertation, University of Alaska, Fairbanks, AK.

Miller, J.M. and Osterkamp, T.E. (1978) The use of Landsat data to minimize flooding risks caused by ice jams in Alaskan rivers, *Proceedings of the Twelfth International Symposium on Remote Sensing of Environment*, Vol. 3, Environmental Research Institute of Michigan, Ann Arbor, MI, pp. 2255–66.

Morisawa, M. (1968) *Streams – their Dynamics and Morphology*, McGraw-Hill New York.

O'Lone, R.G. (1975) Alaskan air support facing challenges. *Aviat. Week Space Technol.* **103**, 14–17.

Osokin, I.M. (1978) Zonation and regime of naleds in Trans-Baikal region, *USSR Contribution to Second International Conference on Permafrost*, 13–28 July, 1973, National Academy of Sciences, Washington, DC, pp. 391–6.

Schmugge, T., Wilheit, T.T., Gloersen, P. *et al.* (1974) Microwave signatures of snow and freshwater ice. In *Advanced Concepts and Techniques in the Study of Snow and Ice Resources* (eds H.S. Santeford and J.L. Smith), US International Hydrological Decade, National Academy of Sciences, Washington, DC, pp. 551–62.

Sellmann, P.V., Brown, J., Lewellen, R.I. *et al.* (1975a) *The Classification and Geomorphic Implications of Thaw Lakes on the Arctic Coastal Plain, Alaska*, US Army Cold Regions Research and Engineering Laboratory, Hanover, NH, RR 344.

Sellmann, P.V., Weeks, W.F. and Campbell, W.J. (1975b) *Use of Side-Looking Airborne Radar to Determine Lake Depth on the Alaskan North Slope*, US Army Cold Regions Research and Engineering Laboratory, Hanover, NH, SR 230.

Strong, A.E. (1974) Great Lakes Temperature Maps by Satellite (IFYGL), *Proceedings of the 17th Conference on Great Lakes Research*, International Association for Great Lakes Research, Ann Arbor, Michigan, pp. 321–33.

Swift, C.T., Harrington, R.F. and Thornton, H.F. (1980a) Airborne Microwave Radiometer remote sensing of lake ice, *EASCON 180 Record*, Proceedings of the IEEE Electronics and Aerospace Systems Convention, 29–30 September and 1 October, 1980, pp. 369–73.

Swift, C.T., Jones, W.L. Jr, Harrington, R.F. *et al.* (1980b) Microwave radar and radiometric remote sensing measurements of lake ice. *Geophys. Res. Letters*, **7**, 243–6.

Wartha, J.H. (1977) *Lake Erie ice: Winter 1975–76*, National Oceanographic and Atmospheric Administration, NOAA Technical Memorandum, NESS 90.

Weeks, W.F., Sellmann P.V. and Campbell, W.J. (1977) Interesting features of radar imagery of ice-covered North Slope lakes. *J. Glaciol.* **18**, 129–36.

Weeks, W.F., Fountain, A.G., Bryan, M.L. and Elachi, C. (1978) Differences in radar return from ice-covered North Slope lakes. *J. Geophys. Res.* **83**, 4069–73.

Weeks, W.F., Gow, A.J. and Schertler, R.J. (1981) *Ground-truth Observations of Ice-covered North Slope Lakes Imaged by Radar*, US Army Cold Regions Research and Engineering Laboratory, Hanover, NH, Report 81–19.

Wiesnet, D.R. (1979) Satellite studies of fresh-water ice movement on Lake Erie. *J. Glaciol.* **24**, 415–26.

Wiesnet, D.R., McGinnis, D.F. and Forsyth, D.G. (1974) Selected satellite data on snow and ice in the Great Lakes Basin 1972–73 (IFYGL). In *Proceedings of the 17th Conference on Great Lakes Research*, Part 1, International Association of Great Lakes Research, Ann Arbor, MI, pp. 334–47.

6

Permafrost

6.1 Hydrological and geological implications of permafrost

Permafrost is a very important condition of the Earth's lithosphere. Permafrost is defined as any material that maintains a temperature below 0°C for at least 2 years. It underlies 20 to 25% of the land surface of the Earth and is also found offshore in many polar areas (Fig. 6.1). In the northern hemisphere permafrost covers an estimated 2.3×10^6 km^2 in area (Fujii and Higuchi, 1978). The occurrence of permafrost is either continuous, or discontinuous in which large masses of unfrozen ground may be interspersed with masses of permafrost.

Permafrost can be dry, consisting of rock and soil, or it can contain a considerable amount of ice. Its thickness is highly variable. Similarly, the thickness of the active layer, the material above the permafrost table that freezes and thaws annually, also varies considerably due to local factors. Permafrost thicknesses of up to 1450 m have been reported in the vicinity of the Markha River in Siberia in the Soviet Union, up to 700 m in the Canadian Arctic and up to 650 m in Prudhoe Bay, Alaska in the United States (Washburn, 1980). The occurrence and thickness of permafrost is dependent upon many factors including climatic history and geomorphic and vegetation conditions, and, especially in the discontinuous zone, slope, aspect, vegetation cover, soil type and snow cover.

Under similar climatic conditions, the thermal conductivity of the rock is especially important in determining the thickness of permafrost (Péwé, 1983). For example, two locations on the North Slope of Alaska: Prudhoe Bay and Barrow have very similar mean annual surface temperatures but permafrost thicknesses vary from approximately 650 m to 400 m. The high thermal conductivity in the sediments in Prudhoe Bay is responsible for the increased

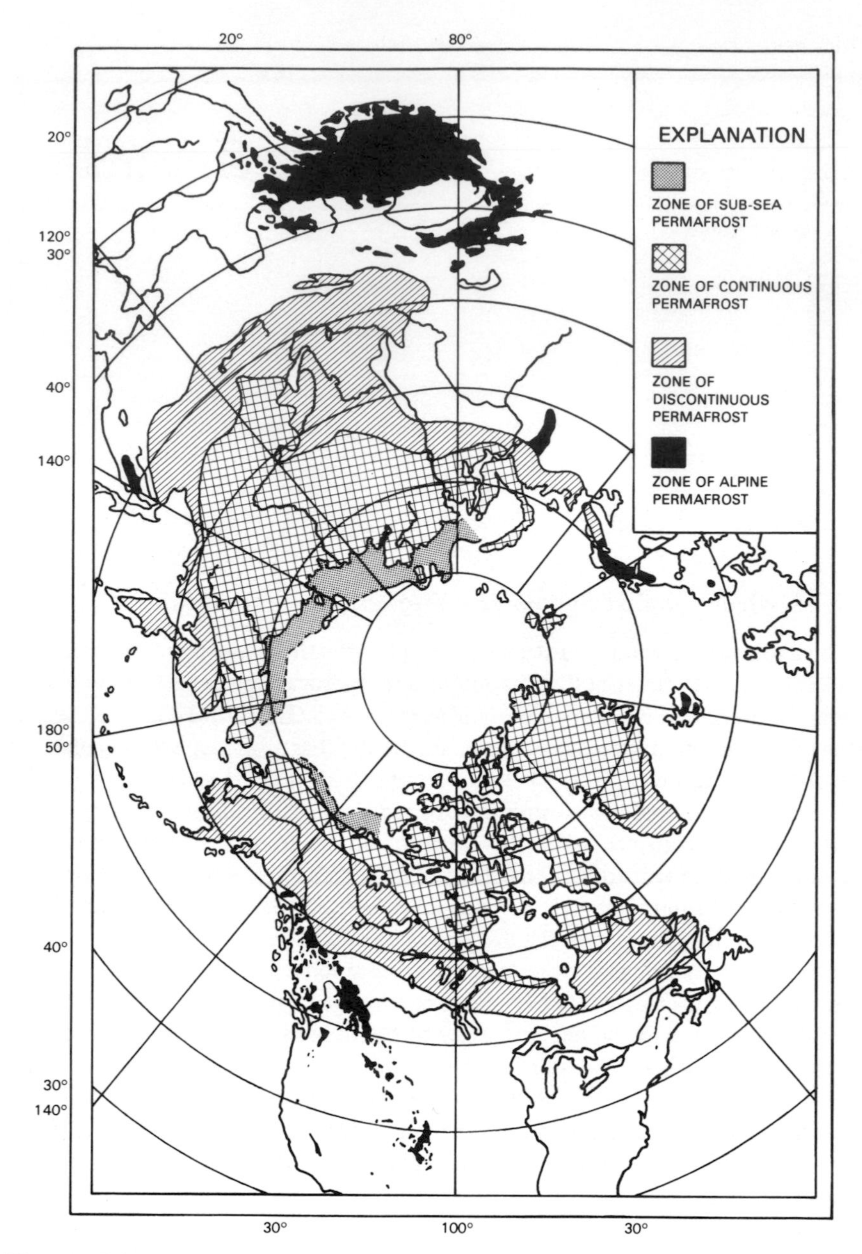

Fig. 6.1 Distribution of permafrost in the Northern Hemisphere (adapted from Péwé, 1983). Reproduced with permission of the Regents of the University of Colorado from *Arctic and Alpine Research*, **15**, (3), 1983.

thickness of permafrost there (Lachenbruch *et al.*, 1982). Rocks having high thermal conductivities will have thicker permafrost and a corresponding decrease in the geothermal gradient (Péwé, 1983).

The variability of permafrost is especially important in discontinuous permafrost regions where human habitation is more likely than in continuous permafrost areas. In a given location, it is important to be able to determine at least the following information if construction is contemplated:

(1) Presence or absence of permafrost
(2) Thickness of the active layer
(3) Ice content of permafrost.

Permafrost is not always in thermal balance with the environment. Present climatic conditions may be warmer than conditions that prevailed at the time when the permafrost formed. In addition, the thermal equilibrium can be easily altered by surface disturbances causing permafrost degradation. Such disturbances can occur naturally (as in the case of fire) or as a result of human activity. Tundra surface disturbances that affect construction and transportation are a result of subsidence of the ground surface. This subsidence is in response to thermal erosion or thawing of ice-rich permafrost.

Ice-rich permafrost contains various forms of ice: ice wedges, ice lenses and pingo ice. Ice wedges are vertically-oriented masses of ground ice formed when surface water (from snowmelt) seeps into thermal contraction cracks in permafrost during spring melt and later freezes (Fig. 6.2). Ice wedges can grow over thousands of years. Ice lenses are smaller forms of ground ice that form parallel to the surface as a result of moisture segregation during the freezing of saturated, fine-grained soils. Pingos are formed both in continuous and discontinuous permafrost areas. They are conical, ice-cored mounds that can grow to over 60 m in height and 600 m in diameter (Embleton and King, 1975). Closed-system pingos are supplied by water from draining lake basins; the water moves upward under cryostatic pressure to form the ice core. Open-system pingos get their source of water from groundwater or springs where artesian pressure develops in taliks (pockets of unfrozen material within permafrost).

Remote sensing is an important method of studying permafrost processes and identifying widespread indicators of permafrost such as vegetation and surficial ice features. Visible, near-infrared and thermal infrared data have supplied important information concerning vegetation and land cover in permafrost terrain, onset and break-up of snow and ice, evapotranspiration, tundra surface disturbances and presence of subsurface ice in some cases. Impulse radars have been successfully employed to obtain quantitative measurements of the active layer and of ice and geological variability within permafrost.

In this chapter, permafrost studies in Alaska are emphasized. Though much work has been done in other permafrost areas such as the USSR, China and

Fig. 6.2 Ice wedge polygons on the Arctic Coastal Plain of Alaska. The ice wedges underlie the narrow troughs which delineate the polygons (photograph by D. Hall).

Canada, northern Alaska represents an area in which numerous remote sensing studies have been performed and are reported upon in the open literature. Thus most of the literature cited is readily available to interested readers.

There are two permafrost related features that have been studied quite thoroughly using remote sensing in northern Alaska: the oriented (thaw) lakes and river aufeis deposits (superimposed ice). Though these features are found in permafrost areas, they are dealt with in Chapter 5 which concerns river and lake ice.

6.2 Vegetation mapping in permafrost areas

Remote sensing, especially aerial photography, has been useful for indirectly detecting the occurrence of permafrost by identification of characteristic landforms and vegetation. Vegetation type and character is generally correlated with the presence or absence of permafrost and/or the thickness of the active layer. Near-surface permafrost provides an impervious layer through which roots and water cannot penetrate (Price, 1972). Thus the presence of permafrost has a profound effect on the nature and type of vegetation present. For example, in central Alaska, it is known that stands of white spruce are generally found in soils that are well drained and/or on south-facing slopes where permafrost is deep or absent. Stands of black spruce generally grow on north-facing slopes and flat areas that are poorly drained and are underlain by permafrost (Viereck and Van Cleve, 1984). Information concerning vegetation, used along with other environmental information, has been employed in environmental models to evaluate the occurrence of permafrost; employment of Landsat Thematic Mapper (TM) and Multispectral Scanner (MSS)-derived land cover data in such models is likely to prove useful to augment other environmental data in the permafrost occurrence models (Morrissey, 1983).

Landsat MSS digital data have been used successfully by several investigators to map vegetation in permafrost areas of northern and central Alaska. Land cover categories, derived from MSS data, are comparable to those which could be derived through the interpretation of far more costly aerial photography (Walker *et al.*, 1982; Gaydos and Witmer, 1983). Walker *et al.* (1982) produced a Landsat-based land cover map legend that may have wide applicability to other areas of the Alaskan Arctic tundra. Land cover classes were mapped in the Arctic National Wildlife Refuge (ANWR) using Landsat MSS digital data. The land cover categories derived for the ANWR are: water, pond-sedge tundra complex, wet sedge tundra, moist/wet sedge prostrate shrub tundra, moist sedge tussock, shrub tundra, partially vegetated areas, barren gravel or rock, wet gravel or mud, and ice. Plate VI shows the results of this work in the ANWR. A thorough field check in the summer of 1982 indicated a high degree of accuracy with this classification scheme.

Using Landsat data, 10 land cover categories were mapped in the 97 000 km^2 National Petroleum Reserve in Alaska (NPRA) by Morrissey and Ennis (1981). The NPRA spans three physiographic provinces as discussed by Wahraftig (1965): the Arctic Coastal Plain, the Arctic Foothills and the Brooks Range. The land cover classification was based on plant communities as delineated using MSS data. The vegetation in this area is largely governed by the polygonal features that are prevalent in areas of continuous permafrost. Table 6.1 shows the relationships among vegetation types, soil, drainage and ice conditions in the NPRA.

Morrissey and Ambrosia (1981) used Landsat data in conjunction with digital terrain data to successfully map forest cover types in an area northwest of Fairbanks, Alaska in the Tanana River basin. Forest cover type can be indicative of the presence of permafrost. Previous studies had shown that MSS data, used alone, could not distinguish among many forest cover types in this area. Topographic position, elevation, slope and aspect are known to influence forest cover type, thus topographic data were combined with Landsat data to achieve an improved classification of forest cover information, specifically species composition, stand size and crown cover density. The use of aerial photography and field verification of MSS-derived land cover categories is essential to the success of any land cover mapping project. The investigator must be very familiar with the study area in order to maintain a high degree of

Table 6.1 Plant communities in relation to landscape elements (after Morrissey and Ennis, 1981)

Community	*Depth of soil thaw*	*Soil moisture*	*Ground ice content*	*Ice wedge polygons*	*Soil type*
Dry mat and cushion tundra	deep	low	low	N/A	Arctic brown
Moist tussock tundra	shallow	low	high	high-centered	Plant tundra
Moist meadow-tussock tundra complex	intermediate	intermediate	intermediate	high and low-centered	Upland tundra
Moist meadow tundra	intermediate	intermediate	low	low-centered	Meadow tundra
Wet meadow tundra	shallow	high	low	low-centered	Meadow tundra

accuracy in the classification. This is especially true in an area of discontinuous permafrost in which abrupt variations in land cover may result from underlying permafrost conditions which can vary widely over a very short distance.

Snow cover is another factor that can be related to permafrost occurrence and/or thickness. A deep snow cover tends to insulate the ground below from the cold while a thin snow cover will not. A thin snow cover that forms early and lasts throughout the winter and into the spring is conducive to the development or perpetuation of thick permafrost and a thin active layer. The high albedo of the snow will reflect solar radiation in the spring thus delaying the warming of the ground. This is the case in northern Alaska and Canada.

6.3 Snow and ice break-up

Satellites provide the overview necessary to analyze onset, duration and deterioration of snow and ice over large areas. Much information concerning the nature, rate and characteristics of snow and ice break-up has been obtained from analysis of satellite data. Analysis of Landsat MSS and NOAA AVHRR data has shown that break-up in the continuous permafrost area of northern Alaska proceeds from the southern foothills of the Brooks Range northward toward the Arctic Ocean, advancing preferentially from lower albedo areas to the higher albedo, undisturbed snow-covered areas. Melting proceeds faster along major rivers when water, flowing downstream, floods the snow cover thus reducing the albedo and increasing absorption of solar radiation (Benson *et al.*, 1975). Other areas of reduced albedo include shallow snow with protruding vegetation and lakes with thin snow overlying the ice (Holmgren *et al.*, 1975). In addition to these natural features, man-made disturbances from oil field camps, roads, etc. have been observed on Landsat MSS data to enhance local ablation (Holmgren *et al.*, 1975). When the snow begins to melt, snow on MSS bands 6 and 7 will have a lower reflectance than snow that contains no surficial water. This is due to increased absorption in these bands by the free water in the surface layers of the snow (see Chapters 3 and 4). Just after break-up in northern Alaska, some snow and ice features such as aufeis (see Chapter 5) remain along the river banks in certain areas of the Brooks Range. On the flat tundra, the surface is very wet after break-up and thus has a low reflectivity on MSS imagery, especially in the near-infrared bands. Drainage is very poor because of lack of relief and the impermeable permafrost underlying the flat coastal plain.

Analysis of sequential Landsat imagery acquired in May and early June revealed that the snow cover in coastal areas lasts several weeks longer than it does in locations that are 50 to 100 km inland due to the influence of the water in the partially ice-covered Arctic Ocean causing lower temperatures to persist longer into summer than in more inland areas (Benson *et al.*, 1975). This effect is influential in the distribution of permafrost and its thickness. Table 6.2 shows mean monthly summer temperature and thawing degree-days for coastal and

Table 6.2 Mean monthly temperatures and thawing degree-days (°C) at coastal and inland stations in northern Alaska during the summer of 1972 and 1973 (after Brown *et al.*, 1975)

	June	*July*	*August*	*Total thawing degree-days*
Barrow (coastal)	0.3 (9)	6.1 (189)	4.8 (148)	346
Barter Island (coastal)	1.2 (36)	4.1 (127)	5.2 (161)	324
Prudhoe (approx. 10 km inland)	2.7 (81)	6.3 (195)	5.7 (176)	452
Happy Valley (approx. 100 km inland)	8.1 (243)	12.6 (390)	9.8 (303)	936

inland stations in northern Alaska. Note the warmer summer temperatures for the inland stations.

Break-up in areas of continuous permafrost, such as northern Alaska, can be especially interesting to observe because the impervious nature of the frozen ground contributes to flooding conditions. Melting of lake ice generally proceeds far more slowly than break-up of river ice. Melting of ice on lakes in northern Alaskan permafrost areas has received much attention and is highly suited to study using visible, near-infrared and radar remote sensors. Even in areas of continuous permafrost, unfrozen water can occur beneath deep (greater than 2 m) lakes and rivers. Such thawed areas can be detected using remote sensing. This work is described in Chapter 5.

6.4 Surface temperature and energy balance studies

By analysis of the annual surface temperature profiles, it appears possible to infer the presence of massive ice within permafrost. It has long been known that aerial photography is useful for detecting the presence of ice-rich permafrost if surface expression in the form of polygonal ground is present. Polygonal ground is indicative of ice wedges below the surface. Le Schack *et al.* (1973) sought to predict massive ice in permafrost in areas that did not have surface polygonal features. Infrared aerial photography at 8–12 and 4.5–5.5 μm was obtained in the Shaw Creek Flats area southeast of Fairbanks, Alaska from an altitude of 230 m. Analysis indicated that polygons were observable on the infrared imagery but not on previous visible aerial photographs of the area. Thus the presence of ice below the polygons is inferred. In addition, they found that the onset of spring thaw caused a differential warming of the subsurface as evidenced in the surface temperature. If an ice lens is present, the rate of increase of the temperature profile at depth will be slow relative to an area with no ice lens present. This will be reflected in the surface temperature.

Calculations showed that during the spring and summer, the surface temperature of the ground above the ice lens continues to lag behind that of the ice-free ground. The maximum surface temperature difference should occur just after the onset of fall.

Thermal infrared sensors were employed by Lougeay (1974) to study glaciated terrain near the Donjek Glacier in the St Elias Mountains, Yukon Territory, Canada. He studied: (1) bare glacier ice, (2) glacier ice overlain by a thin (5–15 cm) layer of morainal material, ice-cored terminal moraine (1–3 m of material overlying an ice core 30 m thick) and morainal material not underlain by ice. Strong thermal contrasts measured at the surface were apparent among the various surfaces (Fig.6.3). Lougeay (1981) also studied Landsat-3 thermal infrared imagery of the Wrangell Mountains in Alaska. Though limited in quantity and quality, the data showed potential for mapping buried glacier ice when compared with ground observations. As seen in Fig. 6.3, thermal contrasts among surfaces vary with time of day because ground surface

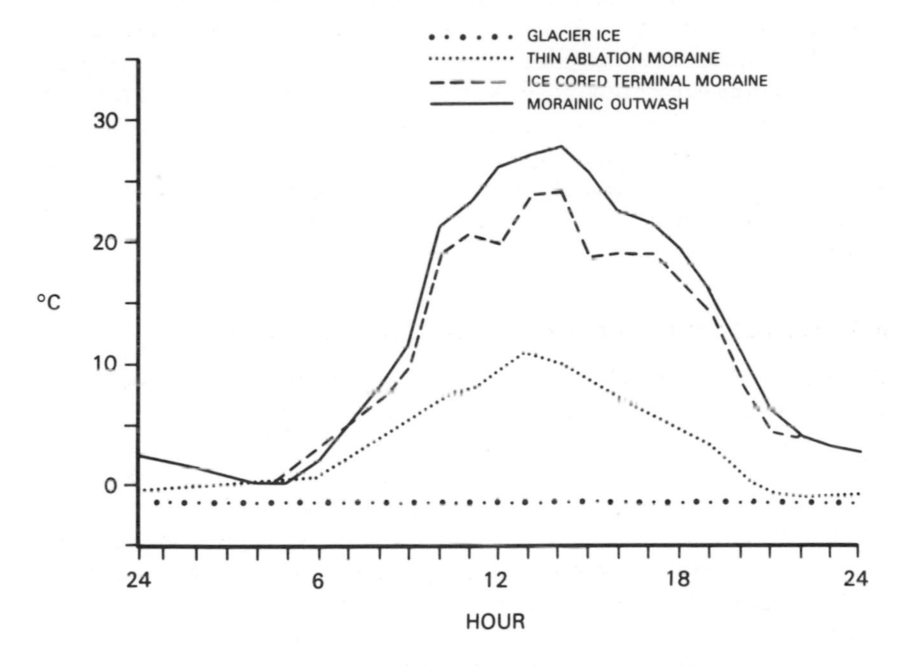

Fig. 6.3 Mean diurnal pattern of radiant temperature under clear skies. Surfaces are: glacier ice, thin ablation moraine, ice-cored terminal moraine, and morainic outwash. The apparent radiant temperature for glacier ice is slightly less than 0°C (Lougeay, 1974). The emissivity of glacier ice is less than 1.0 producing emittance levels which render a radiant temperature of approximately −1°C (Lougeay, written communication, 1984).

materials warm and cool at different rates. In general the greatest thermal contrast among surfaces can be expected in the mid-afternoon (Lougeay, 1974). Though infrared remote sensing represents a potentially useful and interesting method of determining the presence of subsurface ice and analyzing the thermal regime in permafrost areas, more direct methods such as impulse radar sounding (discussed later in this chapter) hold greater promise for quantitative studies.

Thermal infrared surface temperatures measured from satellites have been used for quantitative analysis of the surface energy balance in the Arctic and sub-Arctic. Surface temperatures obtained from the Heat Capacity Mapping Mission (HCMM) and NOAA satellites (see Chapter 2) along with ancillary data can be input to models that permit one to estimate evapotranspiration rates (Gurney and Hall, 1983). This has especially significant implications for permafrost areas in which: (1) little is known about the annual evaporative behavior of the land areas, (2) in-situ data are very difficult to obtain, and (3) the areas involved are extensive.

The energy balance at the surface is governed by:

$$G = R_N + LE + H \tag{6.1}$$

where G is the soil heat flux, R_N is the net radiation flux, and LE and H are the latent and sensible heat fluxes respectively. Remote sensing may be useful for measuring and/or modeling all of the terms, because all terms depend in part on the surface temperature which can be estimated from the thermal infrared data. The latent heat flux gives an estimate of the evapotranspiration from the surface. In addition to the surface temperature and albedo, surface soil moisture and surface roughness (important terms of the equations that solve the energy balance equation) may be estimated using data gathered remotely.

Gurney and Hall (1983) used visible and near-infrared (0.55–1.1 μm) and thermal infrared (10.5–12.5 μm) data from the HCMM satellite at a resolution of 500 m (visible) and 600 m (thermal infrared) to estimate evapotranspiration on 12 May, 1978 in north central Alaska. Conventional meteorological data were used as inputs to an energy balance model constructed to use remotely sensed data. Evapotranspiration within the area studied was found to vary between 2 and 5 mm day^{-1} as shown in Fig. 6.4 which graphically depicts the model relationship found between surface temperature at the time of the HCMM overpass and the daily cumulative evaporation for four typical values of albedo. Such high rates of evaporation are observed in the Arctic for only several weeks after snowmelt each year (Weller and Holmgren, 1974). It is encouraging that the remote sensing data allowed such a good estimate of evapotranspiration to be obtained for this time of the year.

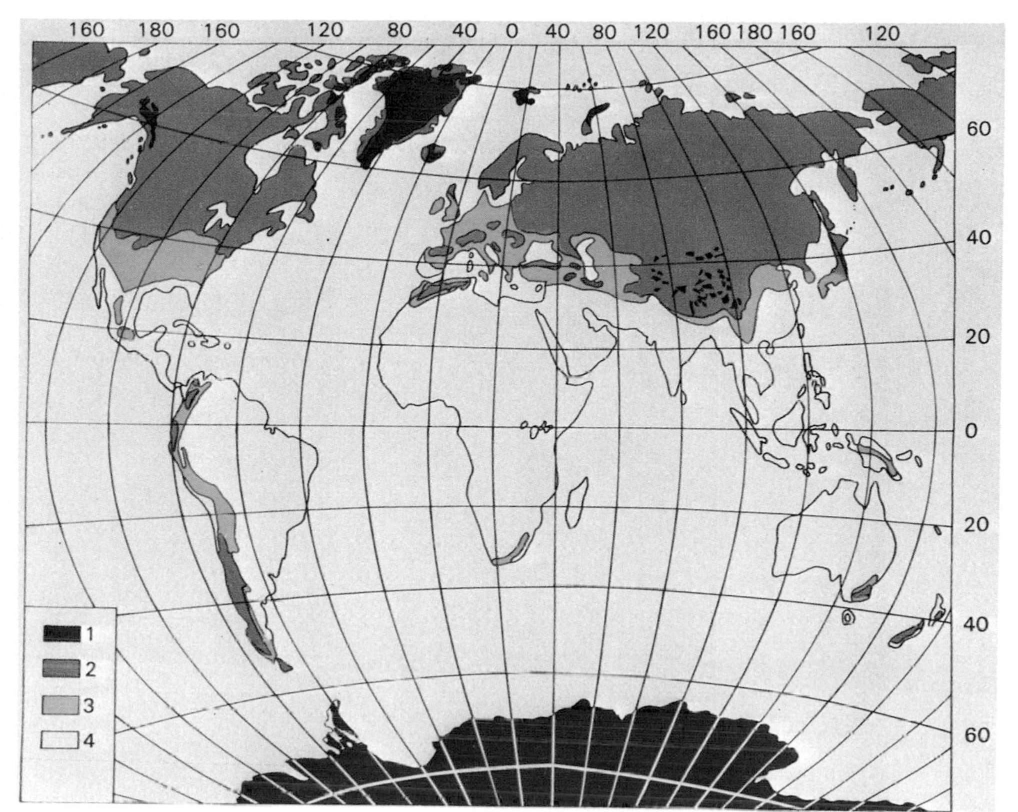

Plate I World distribution of snow cover, drawn after Mellor (1964). 1. Permanent cover of snow and ice. 2. Stable snow cover of varying duration every year. 3. Snow cover forms almost every year but is not stable. 4. No snow cover.

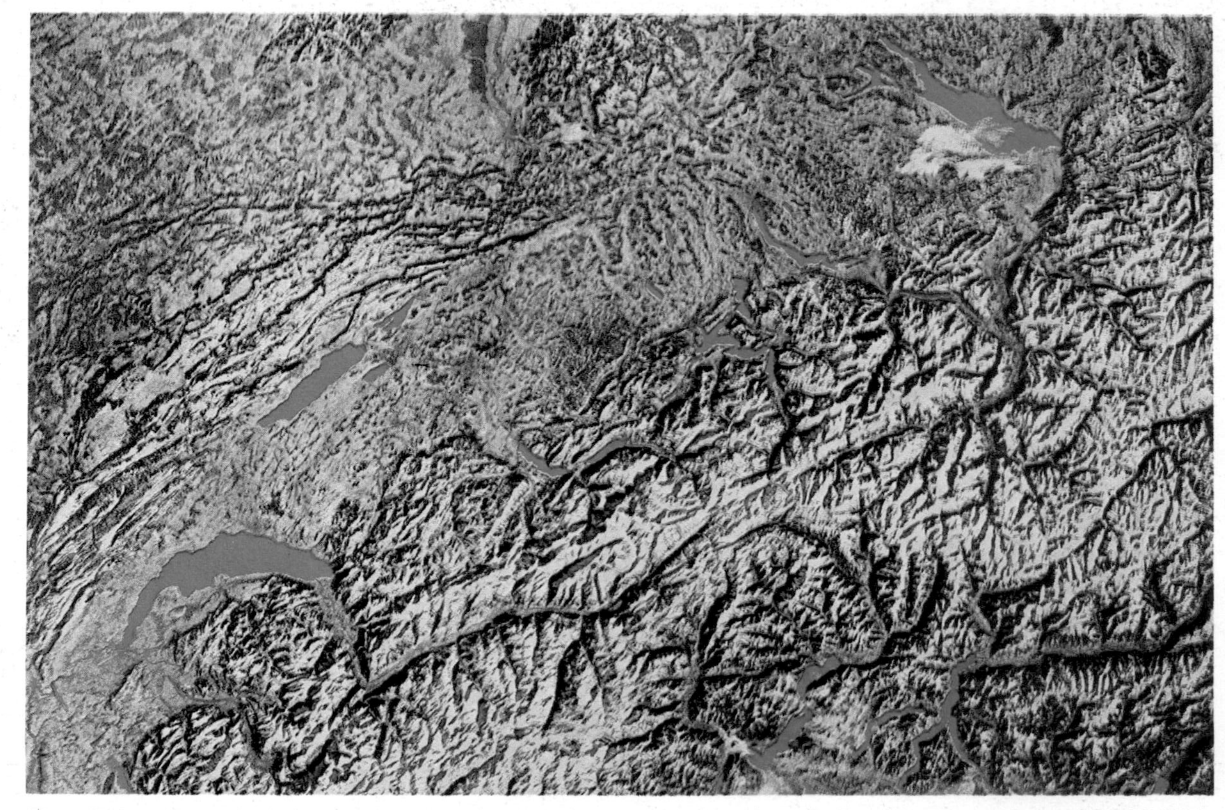

Plate II Snow cover in Switzerland as viewed by Landsat-1 on 1–4 March, 1976. Published by courtesy of NASA/ESA. Image processing by Institute for Communications Techniques, Institute of Technology, Zurich, Switzerland.

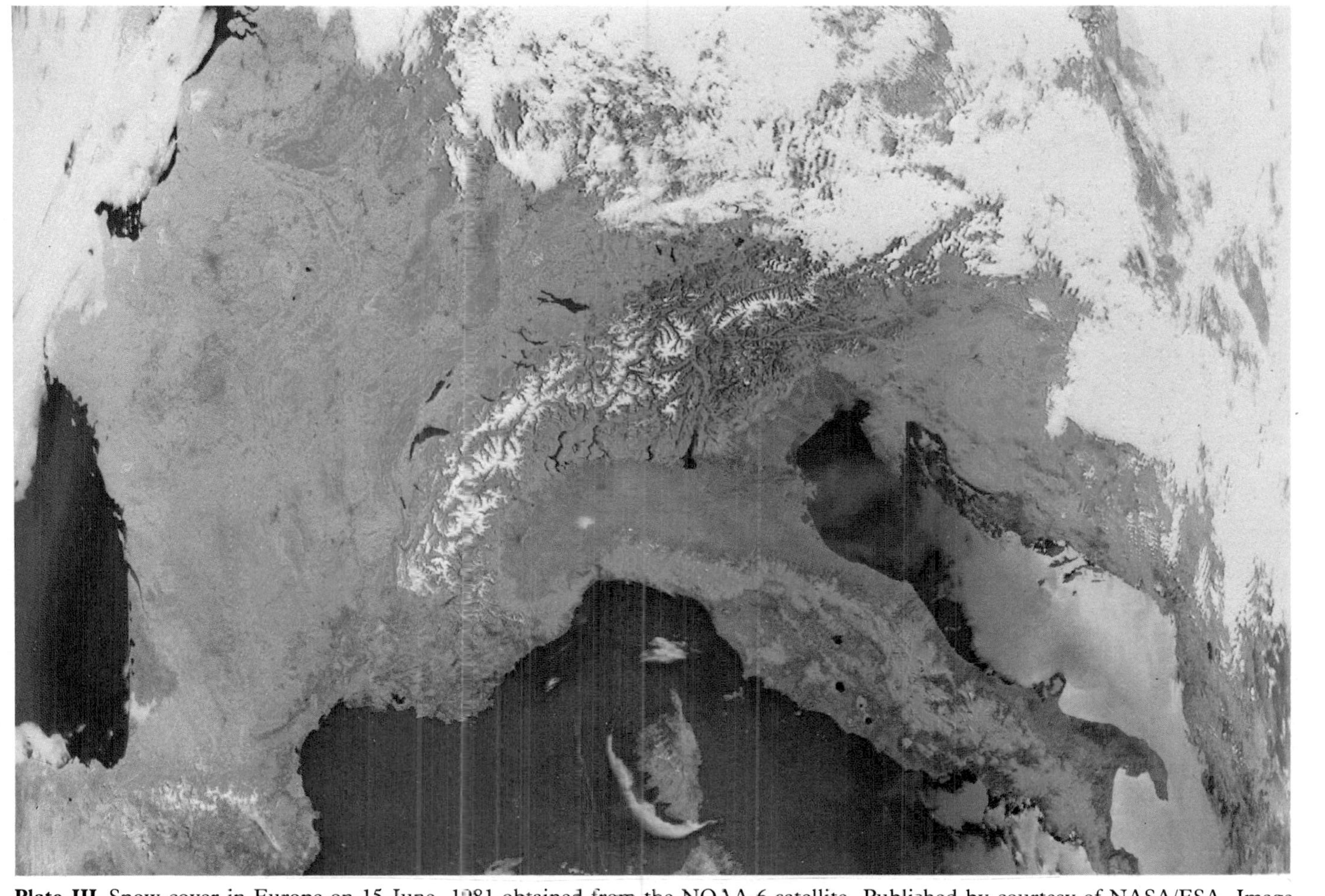

Plate III Snow cover in Europe on 15 June, 1981 obtained from the NOAA-6 satellite. Published by courtesy of NASA/ESA. Image processing by Institute for Communications Techniques, Institute of Technology, Zurich, Switzerland.

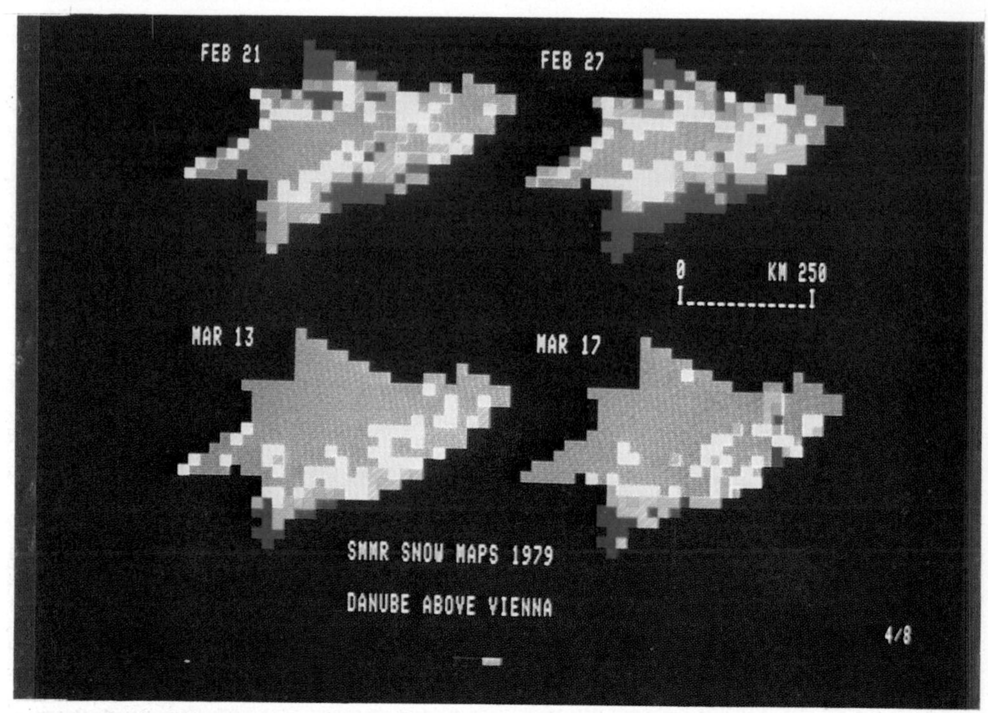

Plate IV Areal extent and depths of the snow cover in the Danube basin above Vienna, Austria, evaluated from Nimbus-7 SMMR data. The color code represents: magenta = snow free; yellow = snow depth 10 cm; green = snow depth 10–20 cm; light and dark blue = snow depth 20 cm. Reproduced by courtesy of Dr Helmut Rott, Institute of Meteorology and Geophysics, Innsbruck, Austria.

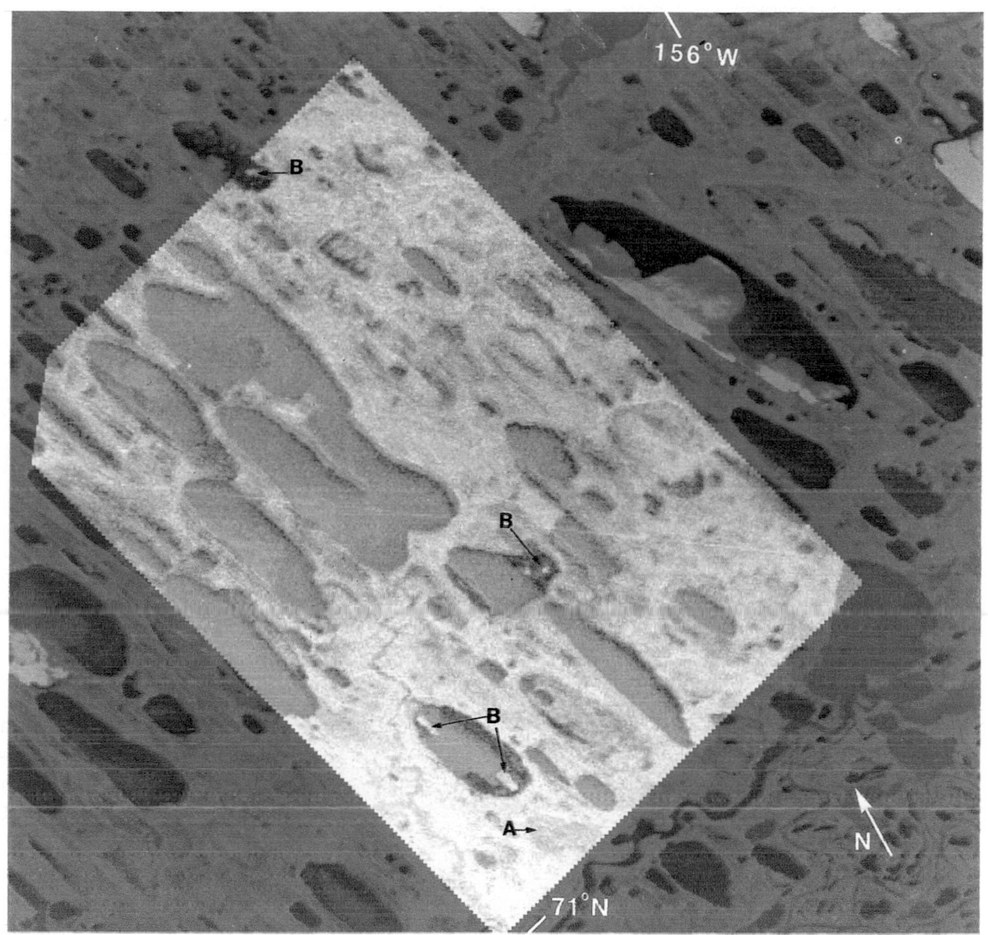

Plate V Seasat SAR data registered to Landsat MSS data (bands 5 and 7) of the oriented lakes south of Dease Inlet: remnant lake basin (A) and emergent vegetation within lake basin (B). Landsat data (Scene I.D. = 30134–21455) were acquired on 17 July, 1978 and Seasat SAR data were acquired on 19 July, 1978. (From Hall and Ormsby, 1983.)

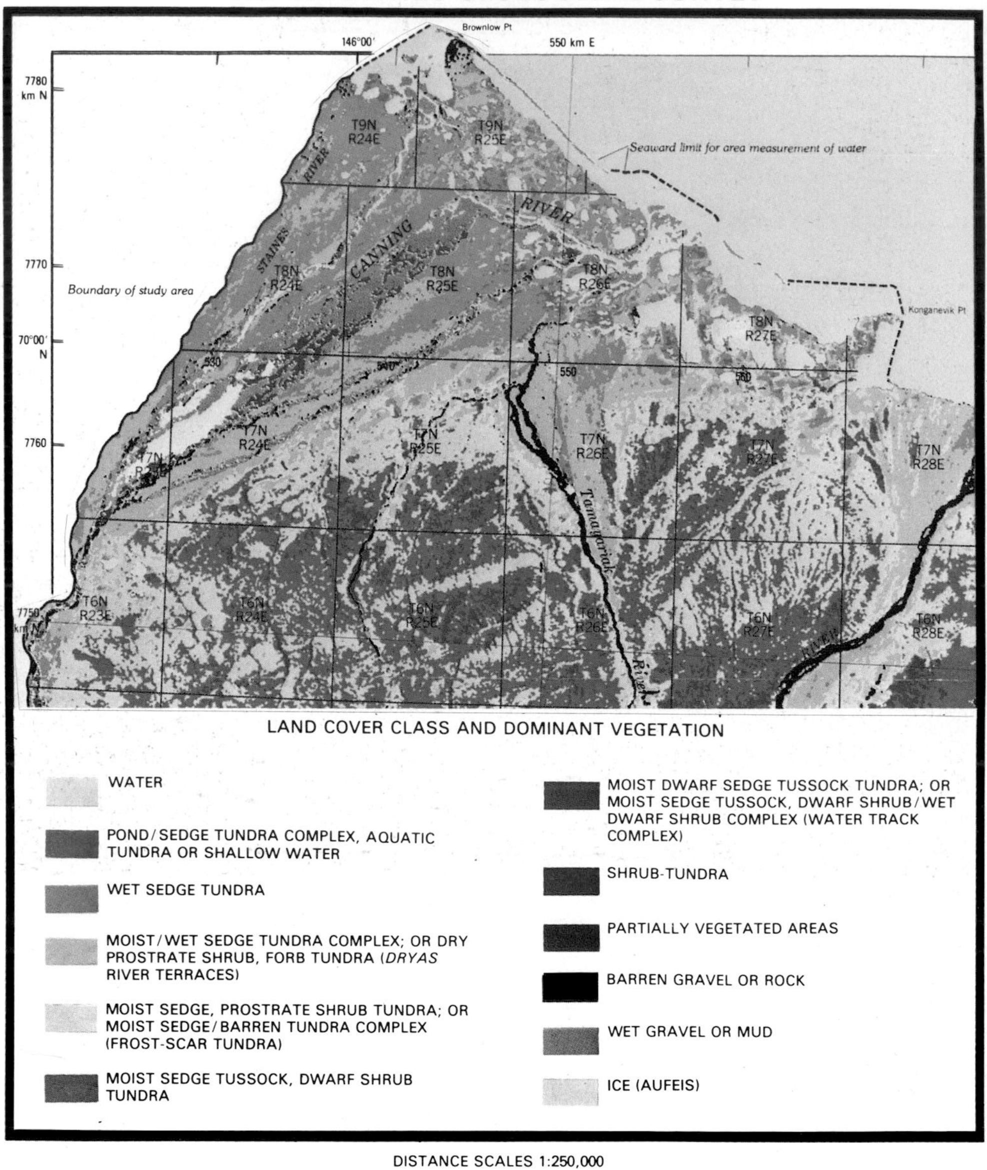

Plate VI Dominant land cover categories in the northern coastal portion of a part of the Arctic National Wildlife Refuge, Alaska, determined from Landsat MSS data (adapted from Walker *et al.*, 1982).

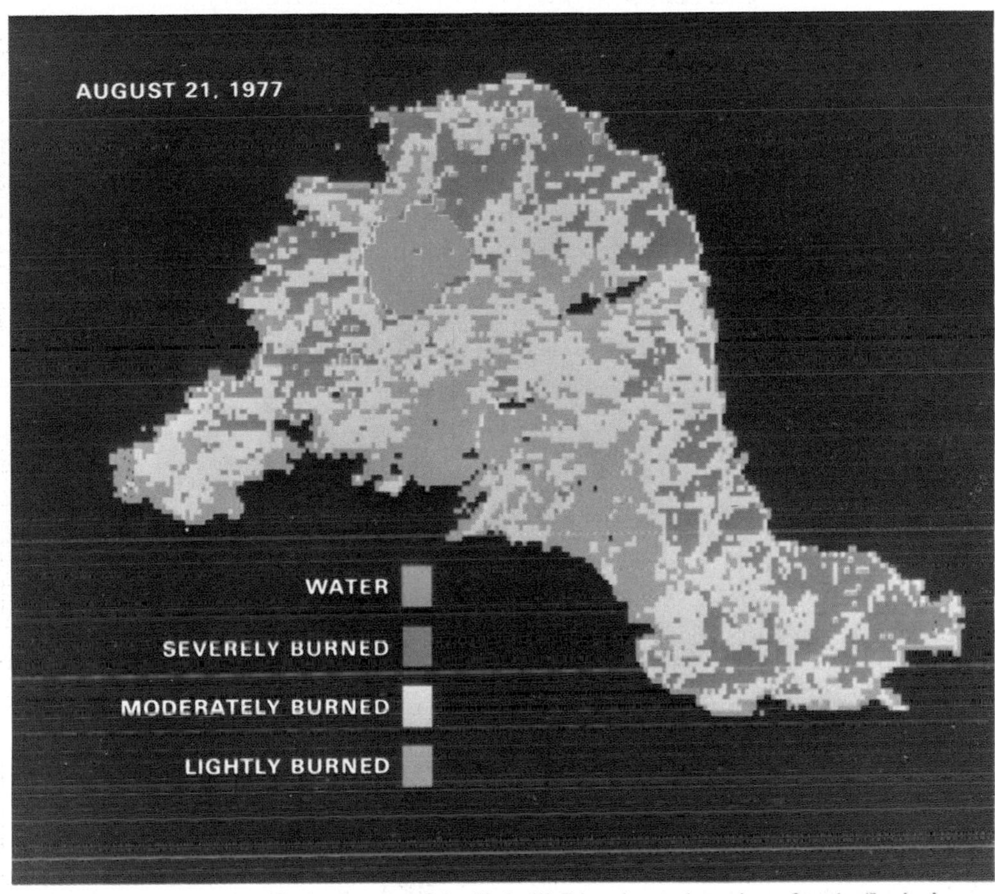

Plate VII Computer classified subscene of the Kokolik River burned area just after the fire in August 1977 (Landsat I.D. 10942–21390, Band 7, 21 August, 1977) (from Hall *et al.*, 1980a).

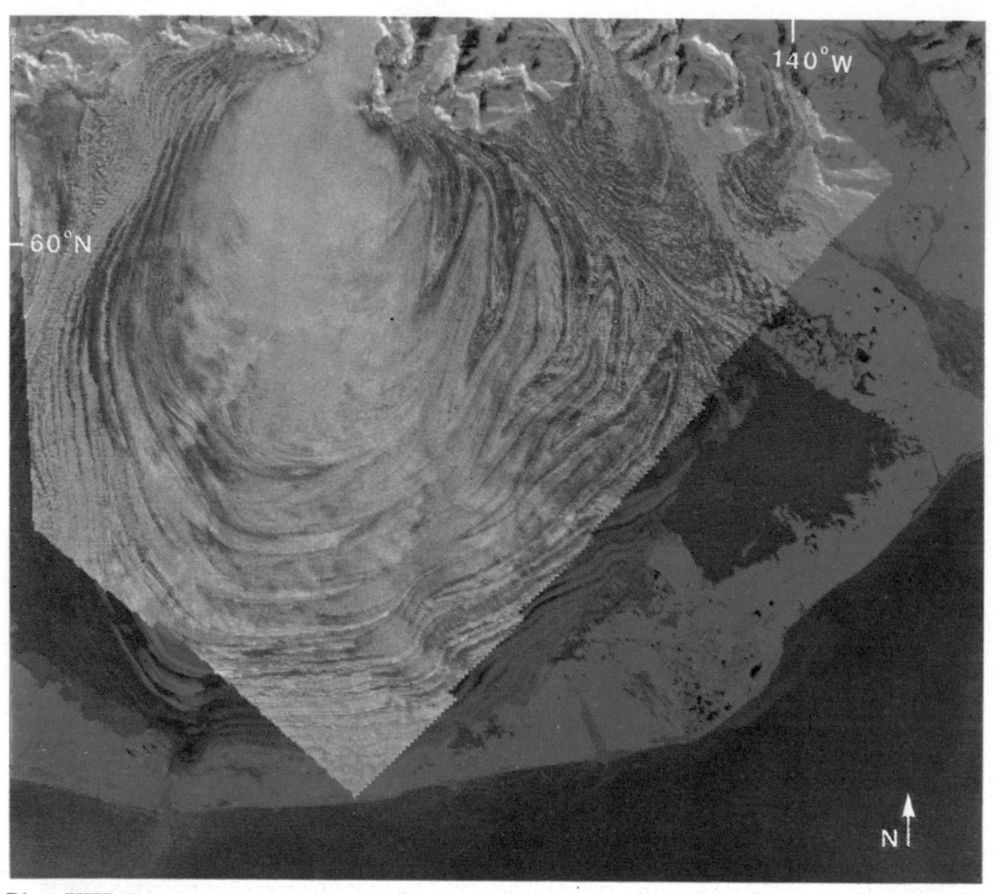

Plate VIII Landsat MSS bands 5 and 7 of the Malaspina Glacier shown in Fig. 7.7, with Seasat SAR data overlaid (from Hall and Ormsby, 1983).

Plate IX False color Landsat (MSS bands 4, 5 and 7) image of Vatnajökull acquired on 22 September, 1973 (I.D. 1426–12070) (from Williams, 1983b).

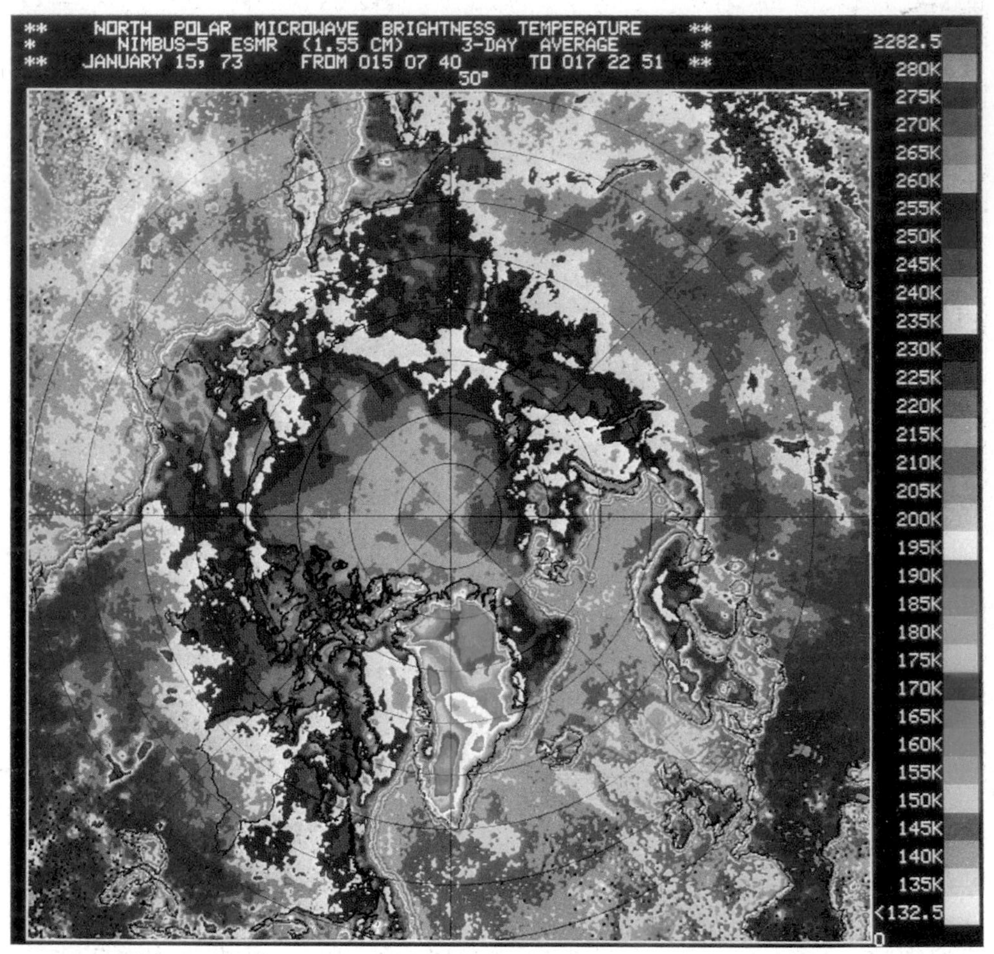

Plate X Nimbus-5 ESMR (λ = 1.55 cm) image of the north polar region showing the Greenland Ice Sheet. Data represent averaged brightness temperatures for mid-January, 1973. (Courtesy of Dr J. Zwally, NASA/GSFC, Greenbelt, MD, USA)

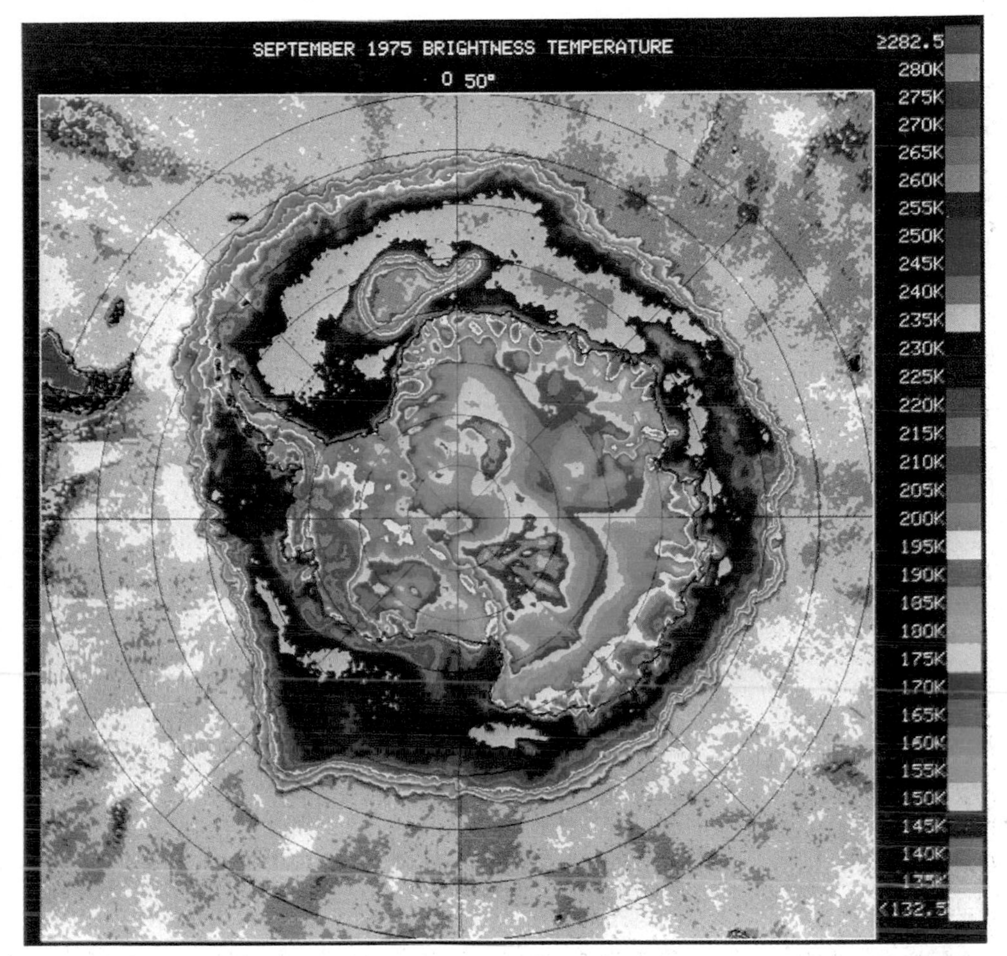

Plate XI Nimbus-5 ESMR (λ = 1.55 cm) image of Antarctica. Data represent averaged brightness temperatures for mid-September, 1975. (Courtesy of Dr J. Zwally, NASA/GSFC, Greenbelt, MD, USA)

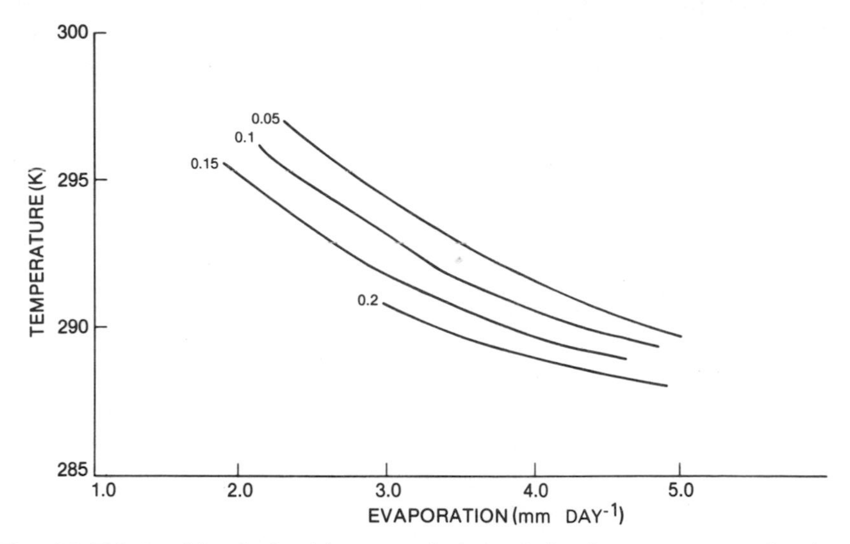

Fig. 6.4 Relationship obtained by numerical simulation between evaporation (mm day^{-1}) and surface temperature (K) for four values of albedo (from Gurney and Hall, 1983).

6.5 Tundra surface disturbances

The tundra is a delicately balanced system that is vulnerable to the vagaries of climate and man. Tundra surface disturbances have been the subject of considerable interest in recent years because of increased public awareness of Arctic areas due to oil discoveries. Remote sensing is useful to aid in the understanding of the characteristics of the permafrost (e.g. vegetation and terrain types), and to monitor and measure changes over time that result from disturbances. Some disturbances to the tundra that have been monitored with Landsat data are: fire, vehicular scars, and activities relating to large construction and exploration projects.

Though fire is an integral part of the natural ecological balance of the tundra, the effects of fire and other disturbances on the tundra are not well understood. The effects of fire on permafrost have been reviewed by Brown and Grave (1979). Fires can have a significant impact on the hydrology of the tundra. If a fire occurs following a prolonged drought, intense burning can result and the permafrost can be stripped of its insulative cover of vegetation. Thawing of near-surface ice can ensue and cause a considerable amount of water to be released and ponded. The result is thermal erosion which will increase each summer until a new thermal equilibrium can be established and the vegetation regenerates. By increasing our understanding of the recovery of tundra following wildfires and other natural disturbances, we can gain insight into the capacity of permafrost for recovery following disruptions caused by man.

When a fire burns in a northern black spruce forest underlain by permafrost as in northern Alaska and Canada, a subsequent increase in active layer thickness occurs. This is not due to the heat of the fire, but to removal of the insulating organic material which lowers the surface albedo and decreases shading effects of the tree canopy (Viereck, 1982). Such changes are potentially possible to monitor with remote sensing.

Forest and tundra fires have been observed on Landsat and NOAA satellite imagery. The NOAA satellite series with its daily coverage provides a good observing platform for viewing fire and smoke patterns over large areas. Ernst and Matson (1977) used NOAA AVHRR data to study smoke patterns in the Seward Peninsula of Alaska on imagery acquired of the extensive burning that occurred during the summer of 1977. Many individual fires and burn 'scars' in northwestern Alaska were observed on both Landsat and NOAA data during and after that summer (Fig. 6.5). Together, NOAA and Landsat MSS digital data can provide quantitative information on reflectance and surface temperature changes in individual burned areas. Temporary and permanent effects on the permafrost must be studied on the ground and inferred from remote sensing.

Reflectances from Landsat MSS digital data were used to monitor the short-term recovery of tundra following one fire in northwestern Alaska that burned in July and August 1977 near the Kokolik River (Hall *et al.*, 1980a). Landsat MSS data were analyzed before, directly after (Plate VII) and one year after the fire. Three burn severity categories were inferred from Landsat spectral response values (digital counts ranging from 0 to 255) and from comparison of the Landsat data with in-situ field measurements. Results showed that one year after the fire, there were significant increases in spectral reflectance within each of the previously determined burn severity categories as seen in Table 6.3. This corroborated measurements on the ground by the US Army Cold Regions Research and Engineering Laboratory (CRREL) in which substantial vegetation regrowth was observed one year after the fire (Johnson and Viereck, 1983). Though vegetation recovery was found to be quite rapid in the first 2–3 years after the fire, recovery slowed dramatically in portions of the burned areas between 3 and 5 years after the fire (Johnson and Viereck, 1983). Recovery of vegetation can be very rapid after a fire in tussock tundra in terms of restoration of rates of primary production; however, the relative abundance of original plant species may not be restored (Fetcher *et al.*, 1984).

In the Kokolik River burned area (Hall *et al.*, 1980a), the Landsat MSS data were interpreted to indicate more complete and faster recovery of vegetation than actually occurred on the ground. This is due to averaging spectral response in the sensor field-of-view which is approximately 80 km^2. This again points out the need for continued ground truth corroboration.

Man-induced disturbances that are sufficiently large are amenable to study using Landsat MSS data. Disturbances resulting from vehicular tracks and

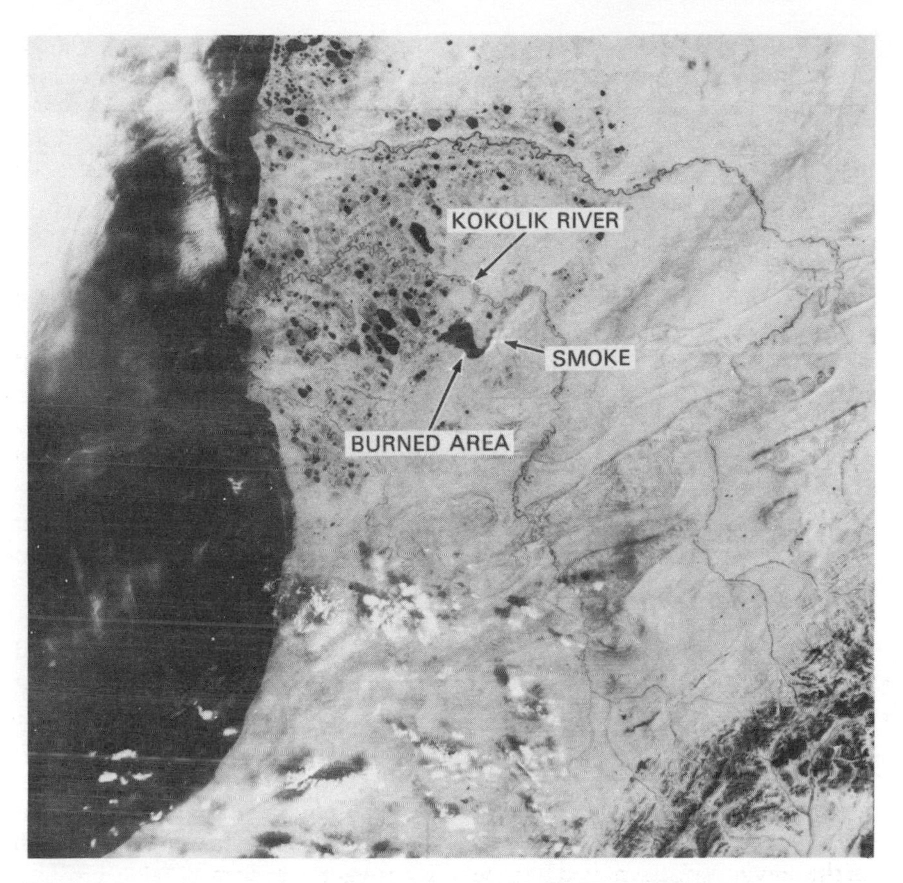

Fig. 6.5 Landsat MSS band 7 image of the Kokolik River burned area 2 August, 1977 (Landsat I.D. 2923–21342). The burned area has a low reflectivity while the smoke is clearly observable with a high reflectivity.

oil exploration activities have caused considerable damage to the tundra. The extent of damage is dependent upon several factors: amount of near-surface ice (water) in the permafrost, time of year of occurrence of the activity, degree of vegetation removal, slope, and type of activity (Brown and Grave, 1979). A considerable amount of vehicular activity caused numerous scars on the tundra in the vicinity of Umiat, Alaska during the 1940s and 1950s. Umiat was the base of operations of the exploration of the Naval Petroleum Reserve in Alaska (now called the National Petroleum Reserve in Alaska, NPRA) between 1945 and 1952. Lathram (1974) was able to identify the largest of the scars near Umiat on enlargements of MSS band 5 Landsat imagery. He concluded that younger scars are more evident on Landsat images than the older scars that have undergone more revegetation and that scars larger than 20 m across could be

Table 6.3 A comparison of percentage of change of ground cover as inferred from spectral response, and as measured in the field, 1977 and 1978 in three categories of burn severity: severely, moderately and lightly burned tundra (Hall *et al.*, 1980a)

Landsat measurements (Mean digital counts, $\bar{x}$)

	1977	*1978*	*Percentage increase*
Severe	8	13	62.5
Moderate	21	30	43.0
Light	30	46	53.0

Ground measurements (percentage of live vascular plant cover)

	1977	*1978*	*Percentage increase*
Severe	5	10	100
Moderate	30	40	33
Light	40	60	50

resolved on Landsat imagery due to the high contrast with the surrounding tundra.

Activities relating to oil exploration in permafrost areas have been closely regulated in recent years and thus damage to the tundra due to overland vehicular transportation has been held to a minimum. Accepted engineering practices that have been developed to work in permafrost areas have proven to be highly successful.

6.6 Subsurface probing of permafrost

Subsurface probing of permafrost by remote sensors is an area of permafrost research that holds a great deal of promise for operational needs. Impulse radars are increasingly being used on airborne platforms for permafrost probing in areas that are very difficult to study from the ground. Airborne impulse radar sounding can be an extremely useful tool for the oil industry: (1) to record the dip of subsurface formations, (2) to gain information on the continuity of subsurface stratigraphy, e.g. faults, and (3) possibly to map the bottom of the permafrost (Unterberger, 1978). Impulse radar is also useful for determining subsurface homogeneity and may therefore be used in conjunction with boreholes to study changes beneath the surface before and after construction. Areas found to be non-homogeneous, using impulse radar data, would require many boreholes in order to determine the precise nature and variability of the subsurface. Homogeneous areas would require few boreholes. The drilling of boreholes is a very expensive procedure and thus minimization of borehole drilling is highly advantageous. Borehole drilling is also disruptive to the ground whereas impulse radar sounding is not.

In addition to its utility for geological studies, impulse radar is also useful for detecting areas of unfrozen water in continuous permafrost areas below frozen lakes and rivers during the winter (Arcone *et al.*, 1979). Numerous investigators have used ground-based and airborne impulse radars to investigate permafrost characteristics: depth to permafrost (thickness of the active layer), taliks and permafrost features such as ice wedges, ice lenses and pingos. Penetration depths of more than 30 m have been observed.

Electromagnetic methods (see Chapter 2) are especially useful for permafrost studies because of the large contrast in dielectric constant that exists between frozen and non-frozen soil. Electrical properties of geological materials in the frequency range 10 to 1000 MHz are strongly dependent on the unfrozen water content of the soil (Annan and Davis, 1978).

Near Tuktoyaktuk, NWT in Canada, Annan and Davis (1976) used an impulse radar with a center frequency of 110 MHz to study structures in permafrost. Their system was mounted on a sled and towed over the ground at a site called Involuted Hill. Involuted Hill consists of a 20 m thick layer of ice overlain by 5 to 10 m of a mixture of clay till, ice and peat. This overlying material also contains ice wedges. The clay till is almost opaque to the radar signal except in the areas where ice wedges are present. The tops of these ice wedges can be 10 to 100 m across and subsurface data can be obtained through these 'windows.' Results of the soundings revealed anomalously low values of the dielectric constant for pure ice ($\epsilon = 2.6$ where it was expected to be $\epsilon = 3.2$). It was determined through analysis of ice cores that a large number of air bubbles existed in the ice causing the dielectric constant to be lower than expected. Thus, the impulse radar data led to an improved understanding of the type of ice present in this area.

Additionally, impulse radar has been able to detect the thickness of the active layer and the existence of ice features in permafrost. Arcone and Delaney (1982) have analyzed ground-based impulse radar data in permafrost areas of Alaska at a time of year when the active layer was thawed to 1 m in depth. They were able to accurately profile the active layer/permafrost interface using this technique. Also using ground-based sounding, Kovacs and Morey (1979) detected the top and bottom of massive ice to a depth of greater than 10 m. Signatures of ice wedges, ice lenses and pingo ice were obtained using an impulse radar system operating from a helicopter at a center frequency of 480 MHz. This particular radar was shown to be useful for detecting and measuring subsurface permafrost features to determine the suitability of the ground for construction and/or transportation activities relating to oil and gas exploration and transportation via pipelines (Hall *et al.*, 1980b). In addition to the geophysical studies, impulse radar sounding has a large potential as a non-invasive monitoring tool for analysis of the condition of underground pipelines and other structures in permafrost and non-permafrost areas.

References

Annan, A.P. and Davis, J.L. (1976) Impulse radar sounding in permafrost. *Radio Sci.* **11**, 383–94.

Annan, A.P. and Davis, J.L. (1978) High frequency electrical methods for the detection of freeze–thaw interfaces, *Proceedings of the Third International Conference on Permafrost*, National Research Council of Canada, Ottawa, Ontario, pp. 495–500.

Arcone, S.A., Delaney A.J. and Sellman, P.V. (1979) *Detection of Arctic Water Supplies with Geophysical Techniques*, US Army Cold Regions Research and Engineering Laboratory, Hanover, NH, CRREL Report 79–15.

Arcone, S.A. and Delaney, A.J. (1982) Electrical properties of frozen ground at VHF near Point Barrow, Alaska. *IEEE Trans. Geosci. Remote Sensing*, **GE-20**, 485–92.

Benson, C., Timmer, R., and Holmgren, B., *et al.* (1975) Observations on the seasonal snow cover and radiation climate at Prudhoe Bay, Alaska during 1972. In *Ecological Investigations of the Tundra Biome in the Prudhoe Bay Region, Alaska*, (ed. J. Brown), Biological papers of the University of Alaska, Fairbanks, AK, Special Report No. 2, pp. 13–50.

Brown, J., Haugen, R.K. and Parrish, S. (1975) Selected climatic and soil thermal characteristics of the Prudhoe Bay region. In *Ecological Investigations of the Tundra Biome in the Prudhoe Bay Region, Alaska* (ed. J. Brown), Biological papers of the University of Alaska, Fairbanks, AK, Special Report No. 2, pp. 3–11.

Brown, J. and Grave, N.A. (1979) Physical and thermal disturbance and protection of permafrost, *Proceedings of the Third International Conference on Permafrost*, Vol. 2, National Research Council of Canada, Ottawa, Ontario, pp. 51–91.

Embleton, C. and King, C.A.M. (1975) *Glacial Geomorphology*, John Wiley, New York.

Ernst, J.A. and Matson, M. (1977) A NOAA-5 view of Alaskan smoke patterns. *Bull. Am. Meteorol. Soc.*, **58**, 1074–6.

Fetcher, N., Beatty, T.F., Mullinax, B. and Winkler, D.S. (1984) Changes in Arctic tussock tundra thirteen years after fire. *Ecology*, **65**, 1332–3.

Fujii, Y. and Higuchi, K. (1978) Distribution of alpine permafrost in the northern hemisphere and its relation to air temperature, *Proceedings of the Third International Conference on Permafrost*, Vol. 1, National Research Council of Canada, Ottawa, Ontario, pp. 366–71.

Gaydos, L. and Witmer, R.E. (1983) Mapping of Arctic land cover utilizing Landsat digital data, *Proceedings of the Fourth International Conference on Permafrost*, National Academy of Sciences, Washington, DC, pp. 343–6.

Gurney, R.J. and Hall, D.K. (1983) Satellite-derived surface energy balance estimates in the Alaskan sub-Arctic. *J. Climate Appl. Meteorol.* **22**, 115–25.

Hall, D.K., Ormsby, J.P., Johnson, L. and Brown, J. (1980a) Landsat digital analysis of the initial recovery of burned tundra at Kokolik River, Alaska. *Remote Sensing Environ.* **10**, 263–72.

Hall, D.K., McCoy, J.E., Cameron, R.M., *et al.* (1980b) Remote sensing of Arctic hydrologic processes, *Proceedings of the Third Colloquium on Planetary Water, Niagara Falls, NY, Partners' Press Inc. pp. 141–9.*

Holmgren, B., Benson, C. and Weller, G. (1975) A study of breakup on the Arctic Slope of Alaska by ground, air and satellite observations. In *Climate of the Arctic* (eds G. Weller and S.A. Bowling), Proceedings of the Twenty-fourth Alaska Science Conference, Fairbanks, AK, pp. 358–66.

Johnson, L. and Viereck, L. (1983) Recovery and active layer changes following a tundra fire in northwestern Alaska, *Proceedings of the Fourth International Conference on Permafrost*, National Academy of Sciences, Washington, DC, pp. 543–7.

Kovacs, A. and Morey, R.M. (1979) Remote detection of massive ice in permafrost along the Alyeska pipeline and the pump station feeder gas pipeline, *Proceedings of the Specialty Conference on Pipelines in Adverse Environments*, American Society of Civil Engineers, New Orleans, pp. 268–79.

Lachenbruch, A.H., Sass, J.H., Marshall, B.V. and Moses, T.H. Jr., (1982) Permafrost, Heat flow and geothermal regime at Prudhoe Bay, Alaska. *J. Geophys. Res.*, **87**, 9301–16.

Lathram, E.H. (1974) Analysis of state of vehicular scars on Arctic tundra, Alaska, *Third Earth Resources Technology Satellite-1 Symposium*, Vol. 1, Section A, National Aeronautics and Space Administration, Washington, DC, pp. 633–41.

LeSchack, L.A., Morse, F.H. and Brinley, W.R. *et al.* (1973) Potential use of airborne dual-channel infrared scanning to detect massive ice in permafrost, *Permafrost–North American Contribution to the Second International Conference*, National Academy of Sciences, Washington DC, pp. 542–9.

Lougeay, R. (1974) Detection of buried glacial and ground ice with thermal infrared remote sensing. In *Advanced Concepts and Techniques in the Study of Snow and Ice Resources* (eds H.S. Santeford and H.L. Smith), National Academy of Sciences, Washington, DC, pp. 487–93.

Lougeay, R. (1981) Potentials of mapping buried glacier ice with Landsat thermal imagery. In *Satellite Hydrology*, (eds M. Deutsch, D.R. Wiesnet and A. Rango). Proceedings of the Fifth Annual William T. Pecora Memorial Symposium on Remote Sensing, 10–15 June, 1979, American Water Resources Association, Minneapolis, MN, pp. 189–92.

Morrissey, L.A. (1983) The utility of remotely sensed data for permafrost studies, *Proceedings of the Fourth International Conference on Permafrost*, National Academy of Sciences, Washington, DC, pp. 872–6.

Morrissey, L.A. and Ambrosia, V.G. (1981) *Tanana Basin Remote Sensing Demonstration Project for Interior Alaska*, National Aeronautics and Space Administration/Ames Research Center, Moffett Field, CA, Final Report.

Morrissey, L.A. and Ennis, R.A. (1981) *Vegetation Mapping of the National Petroleum Reserve in Alaska using Landsat Digital Data*, US Geological Survey Open File Report 81–315.

Péwé, T.L. (1983) Alpine permafrost in the contiguous United States: A review. *Arct. Alp. Res.*, **15**, 145–56.

Price, L.W. (1972) *The Periglacial Environment, Permafrost and Man*, Association of American Geographers Committee on College Geography Resource Paper No. 14.

Unterberger, R.R. (1978) Subsurface dips by radar probing of permafrost, *Proceedings of the Third International Conference on Permafrost*, Vol. 1, National Research Council of Canada, Ottawa, Ontario, pp. 573–9.

Viereck, L.A. (1982) Effects of fire and firelines on active layer thickness and soil temperatures in interior Alaska, *Proceedings of the Fourth Canadian Permafrost Conference, Calgary, Alberta, 2–6 March, 1981*, National Research Council of Canada, Ottawa, Ontario, pp. 123–35.

Viereck, L.A. and Van Cleve, K. (1984) Some aspects of vegetation and temperature relationships in the Alaska Taiga. In *The Potential Effects of Carbon-Dioxide-Induced Climate Changes in Alaska – Proceedings of a Conference* (ed. J.H. McBeath), School of Agriculture and Land Resources Management, University of Alaska, Fairbanks, AK, pp. 129–42.

Wahraftig, C. (1965) *Physiographic Divisions of Alaska*, US Geological Survey Professional paper 482.

Walker, D.A., Acevedo, W. and Everett, K.R. *et al.* (1982) *Landsat-assisted Environmental Mapping in the Arctic National Wildlife Refuge, Alaska*, US Army Cold Regions Research and Engineering Laboratory, Hanover, NH, Report 82–27.

Washburn, A.L. (1980) Permafrost features as evidence of climatic change. *Earth-Sci. Rev.* **15**, 327–402.

Weller, G. and Holmgren, B. (1974) The microclimates of the Arctic tundra. *J. Appl. Meteorol.* **13**, 854–62.

7

Glaciers, ice caps and ice sheets

7.1 Global significance of glaciers

A glacier is an accumulation of ice and snow existing in varying degrees of compaction that moves under its own weight in response to gravitational force. Precipitation in the form of snow adds mass to the system while mass is removed through various ablation processes including surface melting, sublimation and calving of icebergs. In response to these mass inputs and outputs, the motion of the glacier redistributes this mass in an attempt to reach an equilibrium state.

There are several different types of glaciers which are classified according to their size and underlying topography. In this chapter, we will discuss: ice sheets, ice shelves, ice caps, outlet glaciers, piedmont glaciers and valley glaciers. The two ice sheets – in Antarctica and in Greenland – are nearly continuous masses of glacier ice. Ice shelves represent the ungrounded or floating part of an ice sheet that extends into the ocean. Ice caps, far more numerous than ice sheets, are dome-shaped glaciers usually covering a highland area. In Iceland there are several excellent examples of ice caps such as the 8300 km^2 Vatnajökull ice cap. Outlet glaciers flow outward from ice sheets and ice caps through valleys with distinct boundaries consisting of moraines and mountains. Parts of the Antarctic Ice Sheet are drained by ice streams that are broad streams of fast-moving ice surrounded by slower-moving ice. The velocity of outlet glaciers and ice streams can be quite high, i.e. many kilometers per year especially where the terminus ends in a fjord or the open ocean. Valley glaciers are numerous in mountainous regions especially in the higher latitudes. They generally emanate from snow and ice that has been accumulating in a cirque (a bowl-shaped hollow in bedrock) or from an ice field. Piedmont glaciers form when the lobe-shaped expanded terminal portion of a valley or outlet

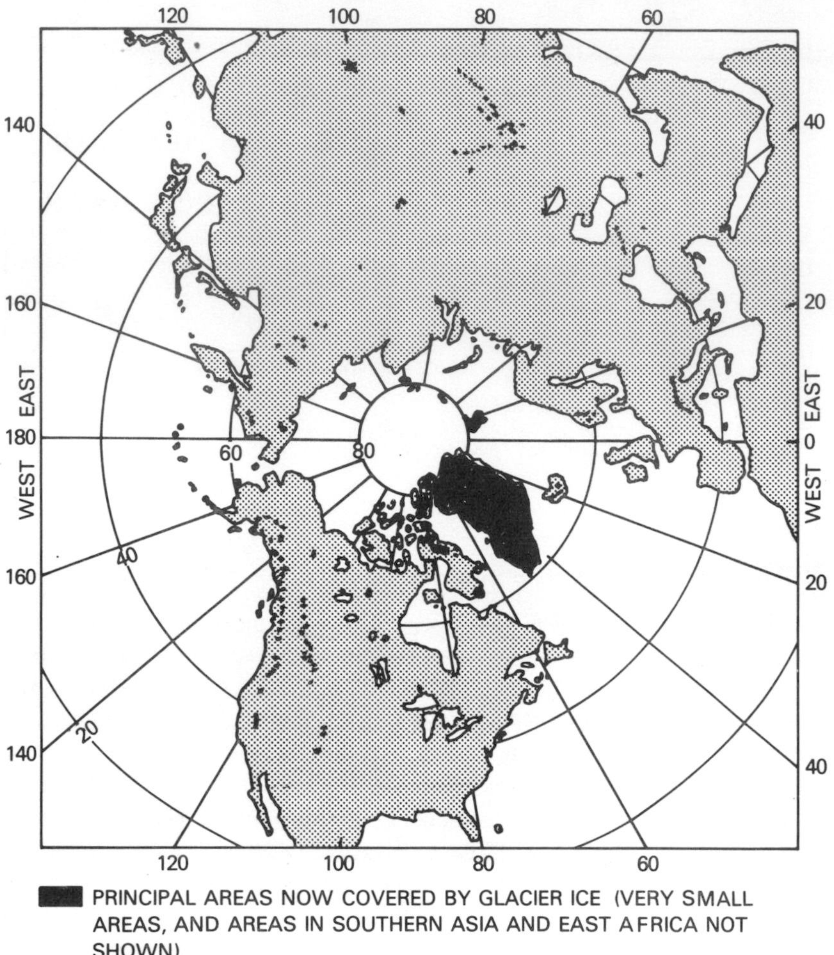

Fig. 7.1 Locations of present-day large glaciers in the northern hemisphere (adapted from Flint, 1971).

glacier spreads out over broad lowlands at the base of mountains. An excellent example of a piedmont glacier is the Malaspina Glacier in southeastern Alaska.

At present, glaciers cover approximately 10% of the Earth's land surface (16 million km^2) and contain 75% of the Earth's fresh water supply. Over geological time, the amount of fresh water stored in glaciers has been highly variable as a result of changing global climate. Sea level varies in response to the amount of water stored in glaciers, and thus glaciers are not only important climatologically, but also have a potential worldwide economic impact. The volume of ice contained in the Antarctic Ice Sheet is sufficient to raise global sea

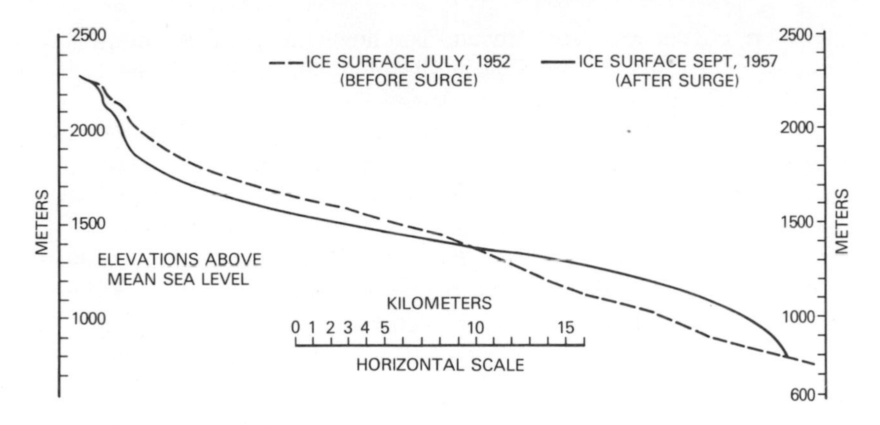

Fig. 7.2 Horizontal profile of Muldrow Glacier, Alaska, before and after a surge (adapted from Post, 1960).

level an average of 70 to 80 m. With the exception of Antarctica, most of the glaciation is concentrated in the Northern Hemisphere (Fig. 7.1). Glaciers can be found on every continent except Australia. Extensive mountain glaciation is found in Alaska, Iceland, Svalbard, Norway, Russian Arctic islands, the Alps, the southern Andes and the Karakoram and Himalayan mountain ranges (Sugden and John, 1976).

Glaciers are economically important as sources of fresh water and for the generation of hydroelectric power in areas of Norway, Iceland, parts of the northwestern United States and the Alps. In addition, iceberg production from coastal glaciers and ice sheets can present interesting challenges with economic ramifications. Icebergs can be extremely hazardous to shipping and thus they must be tracked when they are near shipping routes. Icebergs may also be an important source of fresh water in arid areas if large icebergs can be towed long distances.

In addition to the climatic influence and economic potential, glaciers can exhibit catastrophic behavior. An interesting and poorly understood type of rapid glacier movement is called a surge. A surging glacier is one that experiences a sudden, but short-lived burst of motion after an extended period of quiescence. The glacier becomes heavily crevassed due to the rapid movement of ice in the higher regions into the lower reaches of the glacier. A surge causes the upper portion of the glacier to thin greatly while the lower region thickens (Fig. 7.2). Ice velocities during a surge are typically 100 times higher than velocities during the quiescent phase. For example, on the Medvezhiy Glacier in the Soviet Union (Krimmel *et al.*, 1976), the maximum velocity during a recent surge was reported to be 105 m day^{-1} (Paterson, 1981). Quite often during a surge the terminus will advance many kilometers. The end of the surge phase occurs abruptly and is followed by another long period of quiescence. Thus, surges occur repeatedly and periodically on the same glacier (Post, 1960) with periods as short as 10 years. An important result of the non-steady-state flow characteristic of surging glaciers is contortion of the medial and end moraines. 'Z'-shaped end moraines and looped medial moraines are distinctive signatures of surge-type glaciers which allow them to be differentiated from non-surging glaciers on aerial and satellite imagery.

Another catastrophic type of behavior of glaciers is termed a jökulhlaup. Jökulhlaup is an Icelandic word that means glacier outburst flood. Extensive flooding from jökulhlaups has been reported in Iceland, Norway, the Soviet Union, Alaska, Canada, South America, New Zealand and Switzerland. Such floods in Iceland are produced by two distinct processes: (1) failure of ice-dammed lakes, and (2) creation of large amounts of subglacial water by subglacial geothermal activity and/or subglacial volcanic activity. One of the largest and best-studied lakes subject to repeated jökulhlaups is Grímsvötn, a large (40 km^2) volcanic caldera beneath the western part of the Vatnajökull ice cap in Iceland. The lake slowly increases in volume because of the geothermal activity in the floor of the caldera which melts the overlying ice. The lake is covered by 220 m of ice and drains catastrophically every 5 or 6 years through a complex series of subglacial passages emerging onto the glacier outwash plain to the south (Paterson, 1981). At the peak of a jökulhlaup in 1934 which was also associated with a volcanic eruption, 40 000 to 50 000 m^3 of water sec^{-1} burst from Grímsvötn causing massive flooding (Sugden and John, 1976). A surging glacier can also set the stage for a jökulhlaup if it blocks the flow of a major stream thus impounding water. The failure of an ice-dammed lake will allow the sudden release of the impounded water and will result in flooding.

Glaciers and ice sheets have up to four zones or facies that can be delineated based on snow/ice density and wetness: dry snow zone, percolation zone, wet snow or soaked zone and the ablation zone (Benson, 1962). The dry snow zone is an area that experiences negligible melting. Such areas exist only in the northern interior and higher elevations of the Greenland Ice Sheet, throughout much of the Antarctic Ice Sheet and near the summits of the highest mountains in Alaska and Canada (Paterson, 1981). The percolation zone is so named because localized melting occurs, and water percolates through the surface layers of snow before refreezing. Ice lenses and layers develop beneath the glacier surface. The next zone is the wet snow or soaked zone where all snow deposited since the end of the previous summer becomes wet throughout by the end of the melt season. The fourth zone, the ablation zone, extends from the firn line (the highest elevation to which the snow line recedes during summer) to the terminus of the glacier. The production of meltwater is the main ablation process in this zone. The approximate extent of the facies for the Greenland Ice Sheet is shown in Fig. 7.3.

Remote sensing techniques can be effectively applied to the study of these

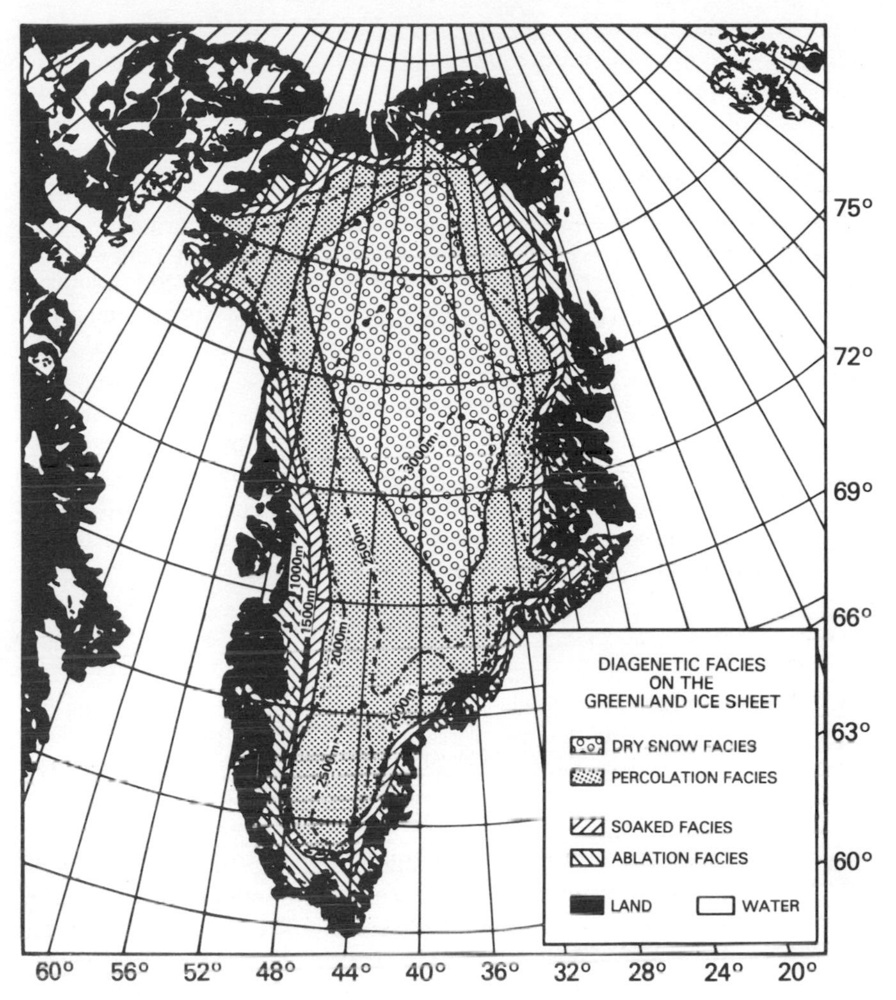

Fig. 7.3 Distribution of the four snow and ice facies on the Greenland Ice Sheet (from Benson, 1962).

fundamental glacier features. The large areal extent and often remote location of glaciers in polar or high mountain areas make remote sensing highly suited to the study of glaciers. It is important to stress that the overwhelming advantage of remote sensing in the study of ice sheets, ice caps and other glaciers is the dual capability of broad coverage and, in the case of microwave sensors, all-weather and all-season data collection. Visible, near-infrared and radar images can be used to monitor the distribution, advance, retreat, and mass balance of valley glaciers, to study surges, jökulhlaups and to track icebergs, while passive microwave data can be used to study the extent and characteristics of the ice sheets including the four snow/ice facies. Radar altimetry and impulse radar sounding can be used to study the elevation and internal characteristics of the ice sheets, respectively.

7.2 Distribution and mass balance of glaciers

The advent of Landsat allowed an enormous amount of information to be obtained concerning the distribution and movement characteristics of the world's glacierized areas. The approximately 80 m resolution of the Landsat MSS and its overlapping orbits in the polar regions make the MSS highly suited to the study of glaciers on a global scale. In fact, a 'Satellite Image Atlas of Glaciers' is being prepared by the US Geological Survey in association with more than 50 glaciologists, and will largely employ Landsat images to show the worldwide distribution of glaciers and to describe their movements (Williams and Ferrigno, 1981). As early as 1973, Meier (1973) recognized the unprecedented value of Landsat data for glacier studies. He was able to identify features on Landsat imagery that were not observable on aerial photography because the features were of such a vast scale as to be unrecognizable on aerial photography. For example, on the Bagley Ice Field in southeastern Alaska, very faint dust bands and medial moraines, seen on Landsat data, revealed previously undetermined directions of ice flow (Meier, 1973).

The firn line or firn limit on a glacier is the position of the snowline at the end of the melt season in late August or early September (in the mid- and high latitudes of the Northern Hemisphere) and can often be seen on Landsat imagery. The position of the firn line is significant, because when observations are made over many years, the average elevation of the firn line can be ascertained. By knowing the average elevation, one can determine whether a glacier had a positive or negative mass balance in a given year by observation of the location of the firn line relative to the average position. For example, in a relatively warm year, in which considerable melting occurs, the firn line will reach a higher elevation than normal. The reverse is true of a colder year.

The equilibrium line is the dividing line between the portion of a glacier that has a net gain of mass over the year and a net mass loss of ice (Paterson, 1981). On most temperate glaciers, the firn line will reach the elevation of the equilibrium line, thus the firn line can be used to approximate the equilibrium line. Knowledge of location of the equilibrium line can be used for mass balance estimates if detailed mass balance studies have been performed on a particular glacier over several years, so that a correlation between specific net mass balance and the firn or equilibrium line can be established (Østrem, 1975). Figure 7.4 is an example of such a correlation on the Nigardsbreen glacier in Norway. If the position of the equilibrium line could be established from Landsat data for any year, one could approximate the specific mass balance of the glacier for that year (in cm water equivalent) by using a map to determine the elevation of the glacier

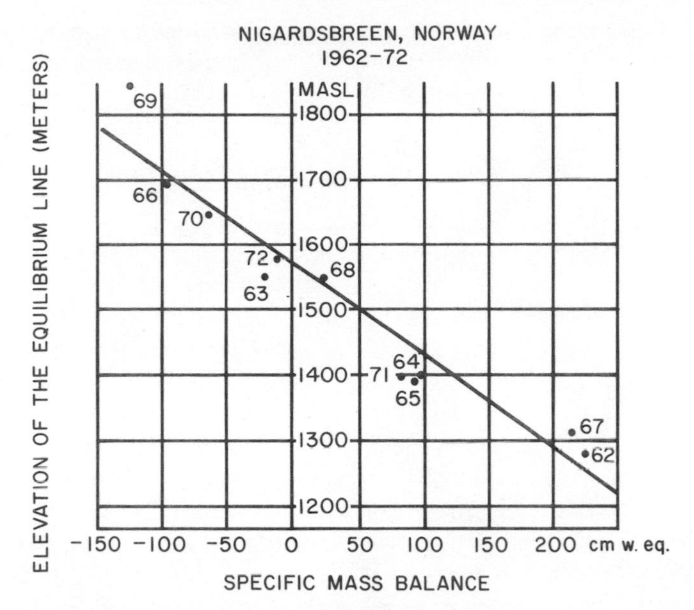

Fig. 7.4 Correlation diagram of the Nigardsbreen Glacier, Norway, showing the relationship between height of the equilibrium line and net mass balance (adapted from Østrem, 1975).

at that equilibrium line position and using Fig. 7.4 to find the corresponding mass balance.

For glaciers in which sufficient data are lacking to determine a relationship between equilibrium line elevation and mass balance, an alternate technique would be to measure the AAR (accumulation area ratio) which is the area of accumulation on a glacier at the end of the summer divided by the area of the entire glacier. An AAR of 0.7 roughly corresponds to a net mass balance of zero (Paterson, 1981). Even on small (less than 10 km^2) glaciers such as the South Cascade Glacier (6.1 km^2) in the State of Washington in the United States, Landsat has adequate resolution to determine the AAR (Meier, 1973).

Figure 7.5 is part of a near-infrared (MSS band 7) Landsat scene acquired on 29 August, 1978 in the Alaska Range in the United States. The firn line (A) on the Ruth Glacier is easily discernable (see section 7.3 for further description of Fig. 7.5). A portion of a Seasat SAR subscene of approximately the same area is shown in Fig. 7.6. The demarcation between snow and ice is not visible on the SAR image. This is probably because the L-band wavelength (23.5 cm) penetrated through the upper layers of snow and firn. Clearly, the Landsat MSS is preferred over the Seasat SAR for firn line determination. The Seasat SAR image mainly shows the morphology of the ice beneath the firn – a valuable indicator of ice flow patterns.

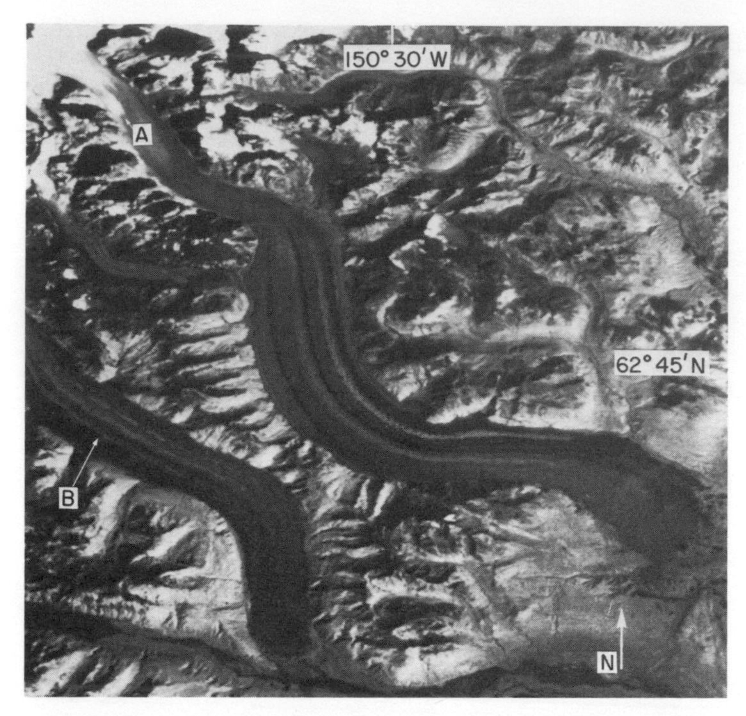

Fig. 7.5 Landsat MSS subscene (band 7) of the Ruth and Tokositna Glaciers, Mt McKinley area, Alaska. Arrow at A points to the firn line on the Ruth Glacier and at B to the wavy medial moraine on the Tokositna Glacier (from Hall and Ormsby, 1983).

The Malaspina Glacier in southeastern Alaska (approximately 2680 km^2 in area) is a large piedmont glacier composed of three lobes fed by numerous valley glaciers in the St Elias Mountains. Comparison of a map and Landsat imagery enabled Krimmel and Meier (1975) to calculate that surface velocities on the Malaspina Glacier varied greatly – from 2 to 300 m $year^{-1}$. The striking morainal patterns, clearly visible on Landsat (Fig. 7.7) and Seasat SAR images, are caused by compressive flow as the glacier moves from higher to lower elevations (Sugden and John, 1976). The wavy patterns and lineaments seen in a digitally enhanced Landsat image of the Malaspina Glacier may relate to bedrock roughness and reflect subglacial relief (Krimmel and Meier, 1975). In an analysis of Landsat MSS and Seasat SAR data of the Malaspina Glacier, Hall and Ormsby (1983) observed that flow lines and morainal patterns were evident on the data from both sensors. It was observed that in the interlobate areas of the Malaspina Glacier (areas A and B on Fig. 7.7) the Seasat SAR and Landsat MSS sensors responded to quite different features. In the SAR image, high returns (light tones) in the interlobate areas are indicative of very rough ice, and flow lines are clearly visible. In the MSS images (both band 5 (Fig. 7.7) and band 7

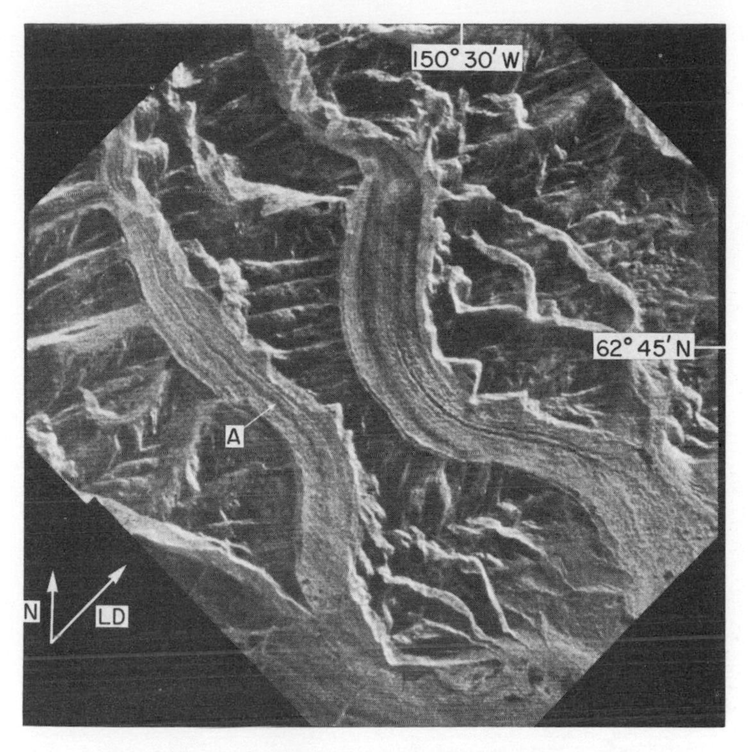

Fig. 7.6 Seasat SAR subscene of the Ruth and Tokositna Glaciers, Alaska. Arrow at A points to the wavy medial moraine of the Tokositna Glacier and LD refers to the look direction of the SAR (from Hall and Ormsby, 1983).

(not shown)) the debris-covered interlobate areas have quite a low reflectance relative to the surrounding ice in the main glacier lobe, known as the Agassiz lobe (C in Fig. 7.7). By viewing the MSS and SAR data side by side or superimposed (Plate VIII), one can see that the interlobate areas are heavily crevassed and covered with glacial debris. The SAR senses the roughness of the debris-covered, crevassed ice and thus has a high return in the interlobate areas; the MSS senses the surficial debris which has a low reflectance in the visible and near-infrared imagery.

The ice in the Agassiz lobe as seen in Plate VIII is smoother and therefore darker in the SAR image compared to the interlobate areas which are quite rugged due to crevassing. Crevasses form when the ice surface is unable to adjust to tensional stresses produced by the ice flow. This is usually caused by rough underlying topography or shear between different regions of ice flowing at different speeds. In either case, crevasse patterns are significant for obtaining a better understanding of ice flow and underlying topography. The interlobate ice on the Malaspina Glacier appears to be heavily crevassed, and therefore it

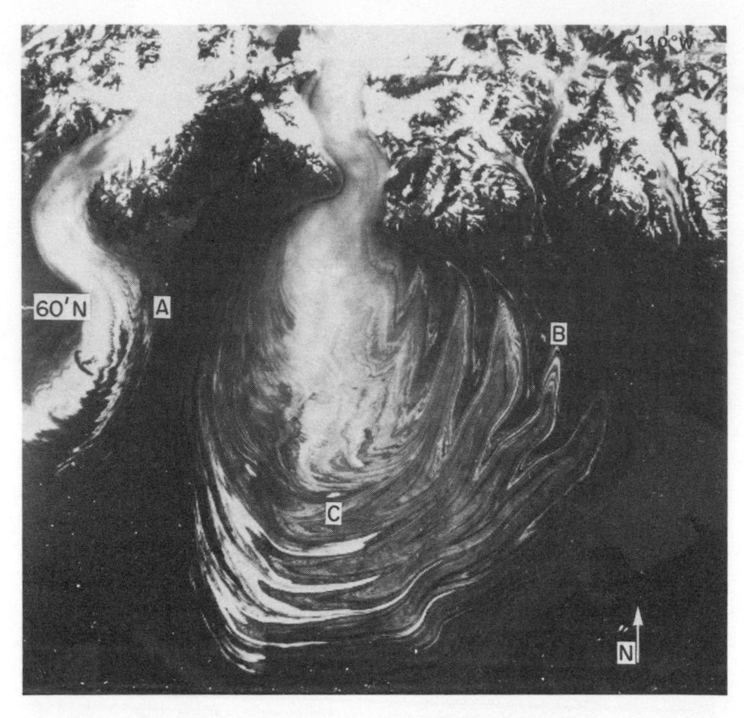

Fig. 7.7 Landsat subscene (MSS band 5) of the Malaspina Glacier, Alaska. A, western interlobate area; B, eastern interlobate area; C, the Agassiz lobe (from Hall and Ormsby, 1983).

must either overlie rather steeply sloping terrain or be caused by shear between adjacent lobes of ice. The intense crevassing may in turn enhance ablation, which may cause more sediment to be released from the ice thus lowering the ice surface reflectance in the visible and near-infrared (Hall and Ormsby, 1983).

A considerable number of remote sensing studies have been done on the glaciers in Iceland. Most of the glaciers in Iceland are ice caps or outlet glaciers (Fig. 7.8) (Williams *et al.*, 1979 and Williams, 1983a). There are a total of 330 named and unnamed ice caps, outlet glaciers and other (smaller) glaciers that can be identified on maps of Iceland at a scale of 1:100 000 or smaller (Williams, 1983b). Landsat imagery has been useful for determining the distribution and extent of Icelandic glaciers. Table 7.1 shows a comparison of Landsat and conventional measurements of the aerial extent of some of the major glaciers in Iceland.

One of the best studied of the Icelandic ice caps is Vatnajökull (Fig. 7.9 and Plate IX). It is the largest glacier in Iceland with an areal extent of approximately 8300 km^2. Analysis of low sun angle (7°) Landsat imagery (Fig. 7.9) revealed

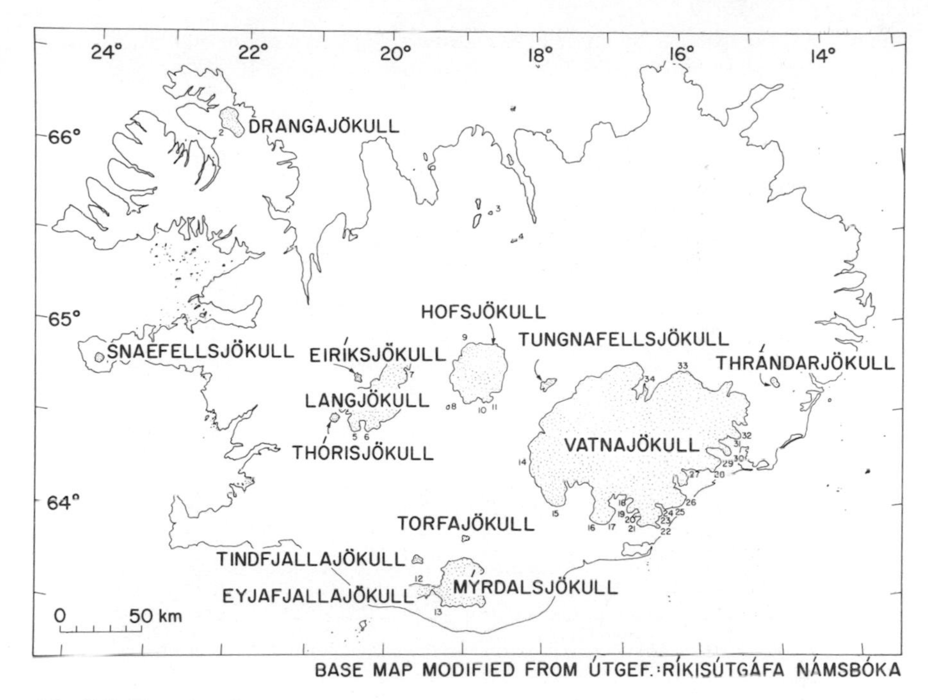

Fig. 7.8 Map showing the principal ice caps of Iceland (from Williams, 1983b).

numerous features on the surface of the ice cap that were not previously observed on aerial photographs. Higher sun angles (>30°) used for standard aerial photographic surveys, and the fact that Landsat imagery offers a regional view not possible on single aerial photographs contribute to the increased information content of the Landsat imagery relative to aerial photography. Figure 7.9 is a Landsat image of Vatnajökull acquired on 31 January, 1973 on which many subtle surface irregularities are visible. Williams *et al.* (1974) and Williams (1976) indicated that some of these irregularities are caused by subglacial topography and others may be caused by subglacial geothermal or volcanic activity. Still others may be due to jökulhlaups (Thorarinsson *et al.*, 1973).

Reflectivity variations on computer-enhanced Landsat images of Vatnajökull (Plate IX) were considered by Williams *et al.* (1979) to correspond to the various snow/ice density zones present on the ice cap (see Section 7.1). An outlet glacier (Merkurjökull) of the Myrdalsjökull ice cap seen on a Landsat image showed a distinct boundary between the soaked zone of the glacier and the percolation zone (Crabtree, 1976). The presence of surface debris and water in the soaked zone of the Myrdalsjökull ice cap in Iceland and along the periphery, tended to disguise the true extent of the ice cap. Because of this, the

Table 7.1 Comparison of areas of the principal glaciers of Iceland (km^2) using conventional and satellite methods (adapted from Williams, 1983b)

Glacier name	*1958**	*Bjornsson†* *(1980)*
Vatnajökull	8538	8300
Langjökull	1022	953
Hofsjökull	966	925
Myrdalsjökull	701	596
Eyjafallajökull	107	78
Tungnafellsjökull	50	48
Thorisjökull	33	32
Thrandarjökull	27	22
Tindfjallajökull	27	19
Eiriksjökull	23	22
Torfajökull	21	15

* Based on Danish Geodetic Institute maps, including post-World War II editions.
† Area calculations made from Landsat MSS 19 August, 1973 (I.D. 1392–12185 and 1392–12191), 22 September, 1973 (I.D. 1426–12070), 9 August, 1978 (30157–11565–D) Landsat images.

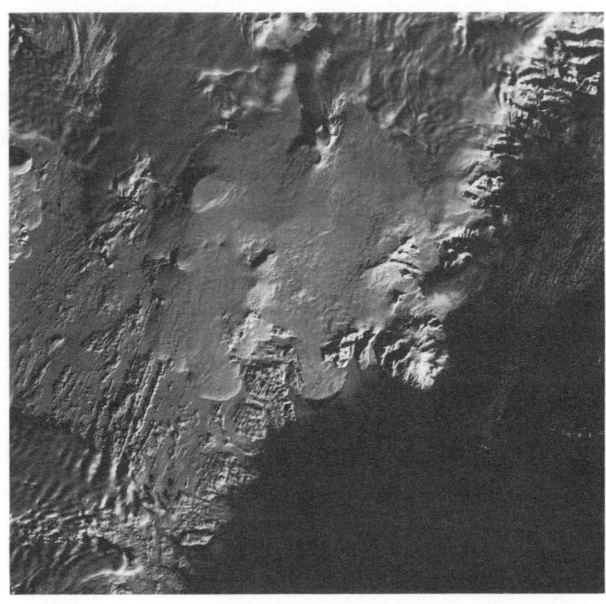

Fig. 7.9 Landsat MSS band 7 image of the Vatnajökull ice cap, Iceland, on 31 January, 1973 (I.D. 1192–12084).

winter scene of Myrdalsjökull gave a better representation of the actual area covered by ice than did summer imagery (Crabtree, 1976). Landsat Thematic Mapper (TM) data with the improved dynamic range (relative to the Landsat MSS sensor) may be employed in the future to delineate the boundaries of the snow/ice density zones on glaciers and ice sheets. Image saturation is generally only severe in TM band 1 in ice and snow covered regions.

7.3 Catastrophic events: surges, jökulhlaups and rapid glacier movement

In addition to being an excellent tool for monitoring the distribution and normal motion of glaciers, Landsat MSS images have been used to locate and study many of the surging glaciers around the world. Landsat has also been useful for studying jökulhlaups that are associated with glaciers and ice caps.

Surging glaciers are normally easy to distinguish from non-surging glaciers on many types of remotely acquired imagery. Landsat MSS imagery has been especially significant to the identification and understanding of surging glaciers because such glaciers are easily visible on Landsat imagery. Wavy, looped moraines (see B on Fig. 7.5) are indicative of a surge-type glacier. Figure 7.5 is a computer-enhanced and enlarged part of a Landsat MSS image of the Tokositna, a surging glacier in the Alaska Range of the United States. The Ruth Glacier, also visible on Fig. 7.5, is not a surge-type glacier. The moraines on the Ruth Glacier are smoother when compared to those on the Tokositna Glacier (Meier, 1976).

The Tweedsmuir Glacier is a surging glacier in British Columbia, Canada that has also been studied using Landsat. A large wave front on Tweedsmuir Glacier marked the limit of intensive crevassing that accompanied its 1973 surge. Though individual crevasses are generally not resolvable on Landsat MSS images, regions of intensive crevassing are discernable because they cause shadowing and thus a lower reflectance than surrounding non-crevassed ice. The wave front on Tweedsmuir Glacier advanced in midglacier at an average rate of 88 m day^{-1} between 15 April, 1973 and 22 July, 1973 for a total advance of about 8.8 km during that period (Post *et al.*, 1976). The down-glacier limit of this intense crevassing is referred to as a shock wave in Fig. 7.10. Williams (1976) measured the motion of Eyjabakkajökull glacier (an outlet glacier of Vatnajökull ice cap) during a surge which began in August 1972. The glacier terminus moved forward 1.8 km as measured between two Landsat scenes acquired approximately 11 months apart – on 14 October, 1972, and 22 September, 1973, respectively (Williams, 1976).

Evidence from aerial photographs and Landsat imagery suggests that there may be a continuum of glacier types. In Alaska, Mayo (1978) found 140 glaciers that he characterized as 'pulsing glaciers.' These pulsing glaciers show some evidence of unstable flow such as wavy moraines. The waves, however, are most

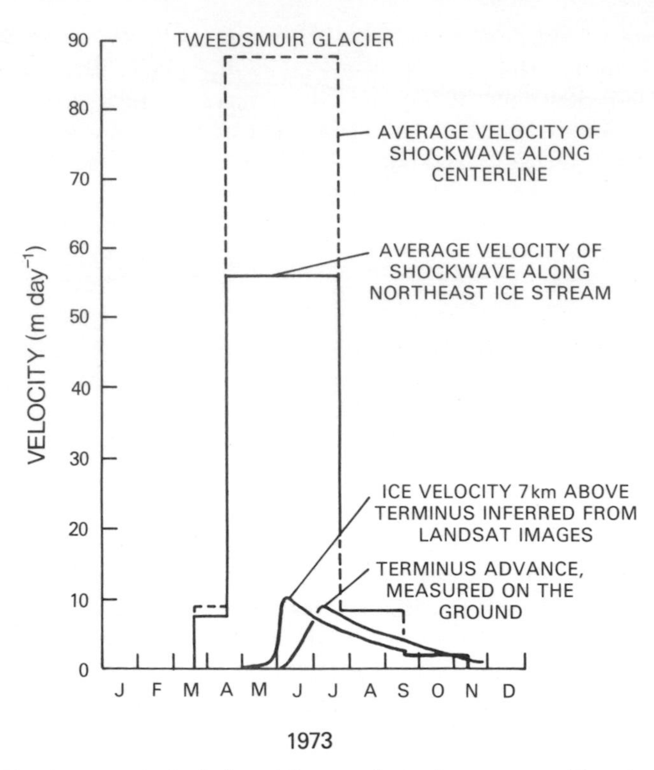

Fig. 7.10 Shock wave velocity inferred from analysis of a sequence of Landsat images of Tweedsmuir Glacier, Alaska, during the 1973 surge (adapted from Post *et al.*, 1976).

pronounced at the glacier terminus and may be present only on restricted parts of the glaciers. Pulsing glaciers are intermediate between surging and non-surging glaciers. As a result of this work on pulsing and surging glaciers, Mayo (1978) asserts that glacier instability may be inherently possible on all glaciers, and that it appears likely that any glacier may become unstable or prone to surging under the proper conditions.

As previously mentioned, a jökulhlaup can be caused by subglacial geothermal and/or volcanic activity or by the failure of ice-dammed lakes. Both types have been identified on Landsat imagery of the Vatnajökull ice cap in Iceland. Grænalón is an ice-dammed lake on the southwestern margin of the Vatnajökull ice cap. Landsat imagery has been used to map the two-dimensional outline and area of the lake (Williams *et al.*, 1974). Grímsvötn, also mentioned previously, has been studied using Landsat imagery. Depressions on the northern rim of the Grímsvötn caldera can be seen on Landsat imagery. These depressions are the result of permanent subglacial geothermal activity and subsidence of overlying ice following withdrawal of meltwater (Thorarinsson *et al.*, 1973). In addition, the recent subglacial course of floodwaters cause surficial depressions that are also visible on Landsat imagery. Analysis of Landsat imagery can give information on the locations of jökulhlaups as well as the course of resulting flood waters (Tomasson, 1975).

Rapid glacier movement can also occur on non-surge-type glaciers. Tidewater glaciers, those that end in tidewater, can rest on shallow shoals of glacial debris in tidewater in their advanced state. Their retreat from the shoals can be quite rapid as was the case with Muir Glacier in Alaska (Field, 1975). At present, the Columbia Glacier, located 40 km west-southwest of Valdez, Alaska is undergoing a drastic retreat (Sikonia and Post, 1979; Meier *et al.*, 1980). This rapid retreat is caused by an accelerated calving of ice resulting in a release of icebergs into shipping lanes. This is the subject of much interest because of the presence of oil tankers in Prince William Sound in Alaska. To date, there have already been frequent instances when the number and size of icebergs near and within shipping lanes have threatened tanker traffic (R. Bindschadler, written communication, 1984). The retreat of the Columbia Glacier is expected to continue, causing increased calving of icebergs into Prince William Sound (Meier *et al.*, 1980). The largest of the icebergs and their paths of movement can be seen and traced on Landsat imagery (see Section 7.5). And the retreat of the glacier can also be monitored using Landsat data, although most of the quantitative studies to date have used photogrammetric techniques on sequential sets of vertical aerial survey photographs (R. Williams, written communication, 1984).

7.4 Greenland and Antarctic ice sheets

Landsat and NOAA satellite images have been used to map the previously ill-defined boundaries of the ice sheets that cover much of Greenland and Antarctica. An atlas of Antarctica is in preparation as a US Geological Survey professional paper entitled 'Satellite Image Atlas of Antarctica,' and will show selected Landsat images of the boundaries of the ice sheet, outlet glaciers, and nunataks (areas free of ice) and will also include tables providing information on the best Landsat images of Antarctica (Williams *et al.*, 1984a). The US Geological Survey (USGS) is also preparing a 1:5 000 000 scale digitally produced image mosaic of Antarctica based on NOAA AVHRR images. Radar altimetry, i.e. from Seasat, is also providing information allowing the quantitative determination of ice shelf edges.

In addition to the very important role of Landsat as a mapping tool, Landsat imagery has been useful for identifying potential areas of accumulations of meteorites which can often be found in some areas of 'blue ice' in Antarctica. Bare glacier ice can also occur near the coast in Antarctica both upstream and downstream of nunataks. The bare ice appears blue because of removal of

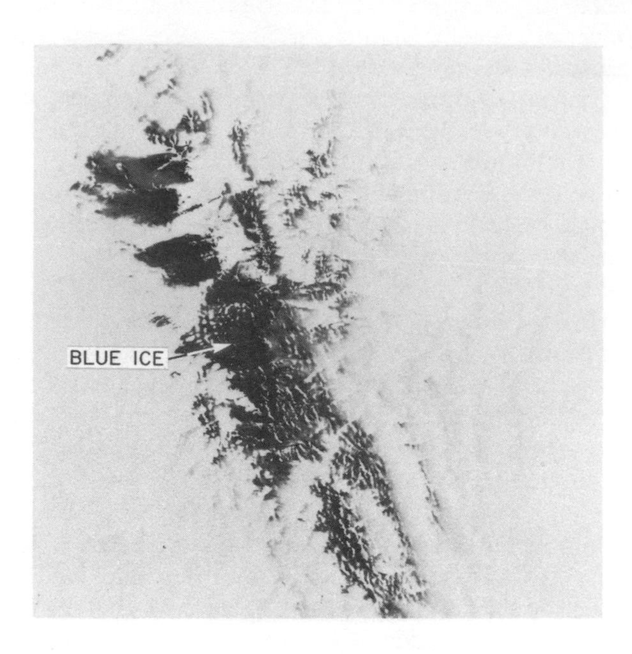

Fig. 7.11 Landsat MSS band 7 image acquired on 2 January, 1974 of the Queen Fabiola Mountains area (Yamato) East Antarctica showing areas of blue ice around nunataks and morainic debris (I.D. 1528–05175) (adapted from Williams *et al.*, 1983).

overlying snow by wind action and by ablation from sublimation, wind-scouring and polishing of the ice surface (Williams *et al.*, 1983). Blue ice areas have a lower reflectivity than surrounding ice on Landsat imagery. Blue ice areas are significant to the collection of meteorites because concentrations of meteorites are found around nunataks or where the normal flow of ice is impeded. These areas correspond to blue ice areas. The greater the ablation, the greater the volume of deposits from within the ice (like meteorites) that would be expected. It is estimated that at least 760 000 meteorites are likely to lie in the East Antarctic Ice Sheet assuming the Earth's meteorite infall rate is one meteorite km^{-2} each million years (Williams *et al.*, 1983). Comparatively few meteorites have been found elsewhere in the world. Figure 7.11 shows a Landsat MSS band 7 image of a large blue ice area in the Queen Fabiola Mountains area of East Antarctica. Landsat MSS and even the poorer resolution NOAA AVHRR data are useful for locating areas of blue ice and thus for planning meteorite searching expeditions (Cassidy *et al.*, 1984; Williams *et al.*, 1984b). Most of the 5300 meteorites found in Antarctica have been found in blue ice areas (Williams *et al.*, 1983).

The spatial resolution offered by the Landsat series and the good coverage offered by the NOAA satellites are useful for the many ice sheet studies as cited, and more. However, a major limitation of the visible, near-infrared and thermal infrared data remains because the sensors cannot image through cloud cover and visible and near-infrared sensors cannot obtain imagery at night. In addition, image saturation in one or more MSS bands often occurs in the polar areas because of the high reflectivity of snow and ice features. Future visible and near-infrared systems should be designed to handle the wide range of reflectivities characteristic of snow and ice-covered terrain (Ferrigno and Williams, 1983).

Since the launch of the Nimbus 5 satellite with an ESMR (Electrically Scanning Microwave Radiometer) on board, data have been obtained of the ice sheets without regard to cloud cover or darkness. The ability of the ESMR to sense through cloud cover is only one trait that makes passive microwave sensors desirable for ice sheet studies. Additionally, the passive microwave data allow information to be obtained, not just of surficial characteristics, such as snow and ice extent, but also of the properties of snow, firn and ice below the surface.

The principles of passive microwave remote sensing and the characteristics of the ESMR are briefly described in Chapters 1 and 2; however, some clarifications are necessary in order to understand the peculiarities of microwave emission from glacier ice. The radiative transfer through the medium must be accounted for, and this renders the basic emissivity (ϵ) equation inadequate:

$$\epsilon = T_B/T_s \tag{7.1}$$

where T_B is the brightness temperature and T_s is the physical temperature of the object. The radiative transfer through the layers of snow, firn and ice which characterize the ice sheets must include the emission, absorption and scattering by a given point within the medium, thus $<T_s>$, an effective physical temperature that is a weighted average of the physical temperature over depth within the medium, is used (Zwally and Gloersen, 1977):

$$\epsilon = T_B/<T_s> \tag{7.2}$$

It is necessary to know the approximate depth from which the radiation emanates. Chang *et al.* (1976) calculated that the radiation emanates from a depth that is 10 to 100 times the length of the wavelength used. Measurement of the amount of subsurface emission and the depth from which it emanates are dependent on the physical characteristics of the ice, firn and/or snow as well as on instrument parameters.

The first Nimbus 5 ESMR images of the Greenland and Antarctic ice sheets revealed large variations in brightness temperature across the ice sheets that did not correlate with physical surface temperature in Greenland and correlated only loosely with surface temperature in Antarctica (Chang *et al.*, 1976).

Observation and modeling have shown that differences in emissivity are related to variations in snow accumulation rates, mean annual temperature, and melting effects on the ice sheets (Zwally and Gloersen, 1977). Specifically, larger grain sizes are known to cause more scattering and to be associated with a lower microwave emissivity. Actively melting wet snow or firn will cause a higher emissivity due to elimination of scattering effects which result from water coating the snow crystals.

Plates X and XI are Nimbus 5 ESMR images of the north and south polar areas showing the Greenland and Antarctic Ice Sheets in January and September, respectively. Note that in northeast Greenland and East Antarctica the brightness temperatures are very low relative to the rest of the ice sheets (Zwally and Gloersen, 1977). Emissivities derived from the physical surface temperature and the T_B for the Greenland Ice Sheet are shown in Fig. 7.12. Calculations have shown that the grain or crystal size is a dominant factor influencing the microwave emission of dry polar firn (Chang *et al.*, 1976). The crystal sizes primarily depend on accumulation rate and mean annual surface temperature. Zwally (1977) estimates that, because of the relationships among crystal size, accumulation rate and surface temperature, the accumulation rate should be measurable to an accuracy of 20% using passive microwave radiometer data. In January 1973 ESMR data, shown in Fig. 7.12, emissivity decreases from the central summit ($\epsilon = 0.85$) to the northern region ($\epsilon = 0.70$) of Greenland which has a lower accumulation rate. In Antarctica, the emissivity (not shown) is low ($\epsilon = 0.70$) over the central plateau in East Antarctica which is also known for low accumulation rates.

There are notable exceptions to the correspondence between accumulation rate and emissivity. In Antarctica, there is a high ($\epsilon = 0.85$) emissivity from the Lambert Glacier (72°S, 68°E) which is not due to a high accumulation rate, but possibly because the glacier has extensive areas of bare or blue ice, exposed by wind action. Low scattering and high emissivity characterize such areas of solid ice. The observed high emissivity may also result from higher surface temperatures in ice streams.

Comiso *et al.* (1982) modeled the seasonal variation of the 1.55 cm microwave emission from stations on the Greenland and Antarctic ice sheets. In their model, depth-dependent parameters were employed in a numerical solution of the radiative transfer equation. Stations were selected for which snow grain sizes had been measured as a function of depth. Results for two stations in Antarctica: South Ice and Plateau, are shown in Fig. 7.13. In general, the agreement between the calculated and observed brightness temperatures was good at both stations. The seasonal variation in T_B is strongly dependent upon emission depth.

The visible, near-infrared and passive microwave imagery are extremely useful for determining annual and interannual changes and other ice sheet characteristics; however, none of the aforementioned techniques can permit the

Fig. 7.12 Contours of constant emissivity on the Greenland Ice Sheet on 11 January, 1973. Emissivity values were obtained by taking the ratio of the brightness temperatures obtained from Nimbus 5 ESMR (λ = 1.55 cm) data to those obtained from infrared instruments (λ = 10 μm) also on-board the Nimbus 5 satellite (from Chang *et al.*, 1976).

calculation of mass balance. Radar altimetry shows much promise in this regard. The mass balances of the Greenland and Antarctic ice sheets are unknown. Baseline information of the elevation of the surface of the ice sheets is necessary to determine if the ice sheets are thickening or thinning. It is not feasible to determine the mass balances of the ice sheets by conventional surveying or direct measurement (Brooks *et al.*, 1978). Satellite radar altimetry provides a means of determining the elevation of the ice sheets for determination of ice thickness changes and thus mass balance (Zwally *et al.*, 1983). Ice thickening would be indicative of a positive mass balance and ice thinning would indicate a negative mass balance. In addition, large-scale ice surface morphology can be indicative of basal topography and ice flow direction. This can also be detected by radar altimetry. Ice surface elevations are

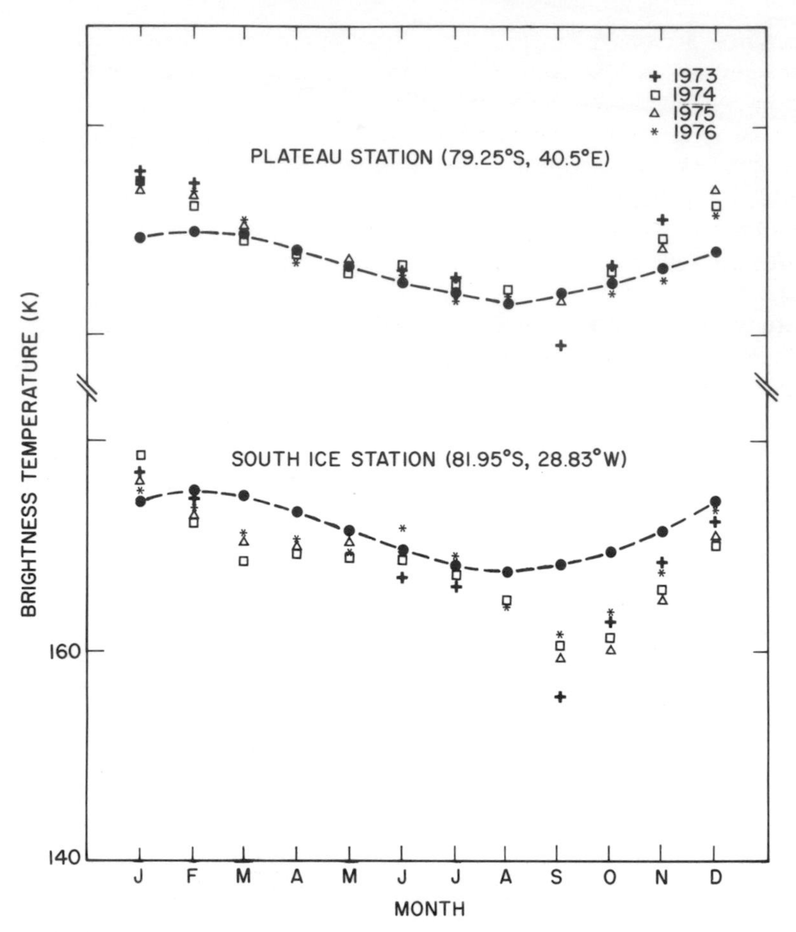

Fig. 7.13 Observed and modeled seasonal variation of T_Bs at Plateau and South Ice stations, Antarctica. Model results (dashed line) are compared with 4 years of Nimbus 5 ESMR ($\lambda = 1.55$ cm) T_B data (from Comiso *et al.*, 1982).

computed by subtracting the altimeter height measurements (the measured time between transmission and return pulses) from the calculated satellite altitudes above a reference ellipsoid of the Earth, and also subtracting the geoid height (local sea level height above the ellipsoid) (Brooks *et al.*, 1978). Range errors include: inaccuracies in orbit determination, variations in atmospheric path length and variations in the timing of the leading edge of pulses returned to the satellite (Brenner *et al.*, 1983).

The Seasat radar altimeter (see Chapter 2) and the radar altimeter on-board the NOAA GOES satellite (a 13.9 GHz altimeter designed to measure ocean wave height) have been successfully used to obtain surface elevations of the ice sheets to an accuracy of approximately 2 m. Several corrections must be applied to the raw data as described for the Seasat altimeter by Brenner *et al.* (1983) and Zwally *et al.* (1983) to achieve this accuracy. A detailed contour map at 50 m elevation intervals was produced from Seasat radar altimeter data for the central Greenland plateau (Zwally *et al.*, 1983). This is shown in Fig. 7.14. A similarly produced topographic map with 100 m contour intervals of a portion of Antarctica is shown in Fig. 7.15. Figure 7.16 shows an exaggerated surface profile across the Greenland Ice Sheet derived from GOES 3 radar altimeter data.

Study of the mass balance of the individual drainage basins within the Greenland Ice Sheet is one way to assess mass balance and this method appears feasible (Bindschadler, 1984). Delineation of drainage basins can be accomplished with precise knowledge of the surface elevations as determined from radar altimetry. Bindschadler (1984) delineated many of the major drainage basins of the Greenland Ice Sheet using Seasat radar altimeter data. In

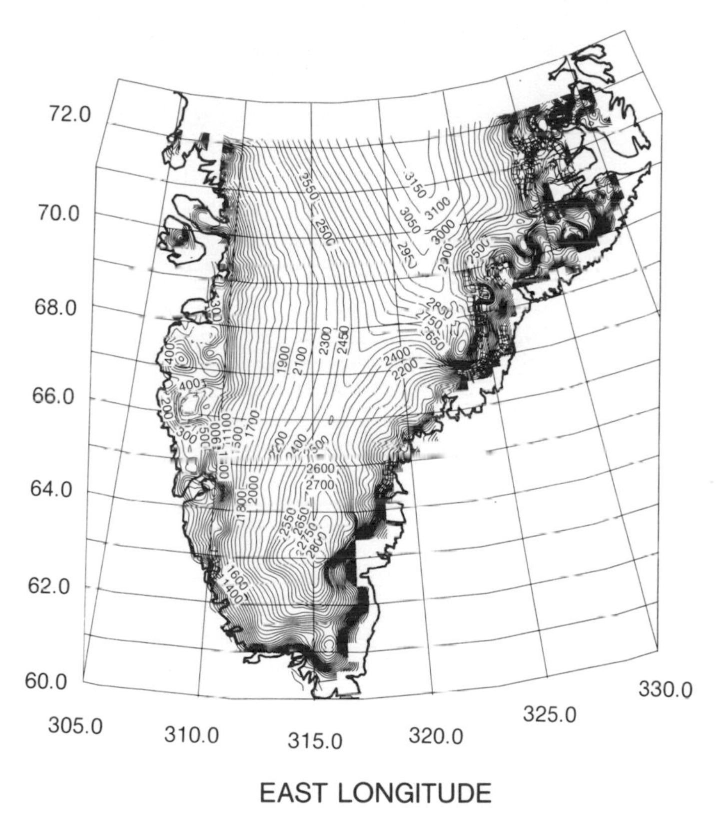

Fig. 7.14 Topographic map (50 m contour interval) of part of the Greenland Ice Sheet derived from Seasat radar altimeter data (from Zwally *et al.*, 1983).

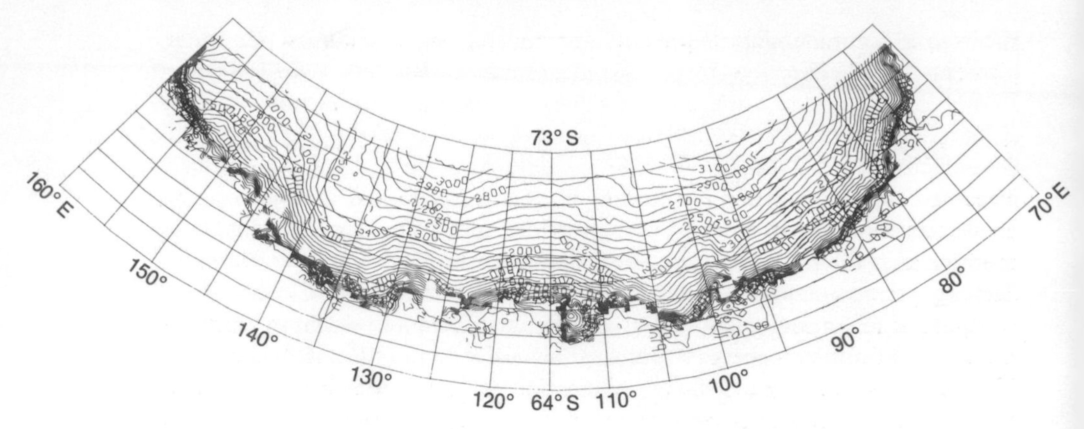

Fig. 7.15 Topographic map (100 m contour interval) of part of Antarctica derived from Seasat radar altimeter data (from Zwally *et al.*, 1983).

particular, he studied the basin drained by the Jakobshavns Glacier in western Greenland. Jakobshavns Glacier drains a significant portion of the ice sheet area – between 3.7 and 5.8%, thus any widespread change in the mass balance of the ice sheet would be reflected in a change of flow of the Jakobshavns Glacier (Bindschadler, 1984). It is believed that the Jakobshavns Glacier is in approximate equilibrium. Detailed studies of the other drainage basins are indicated.

The practicality of radar altimetry for use in determining mass balance of the ice sheets has been demonstrated. Regular ice sheet elevation measurements at intervals of 2 to 5 years are needed before accurate mass balance estimates can be attempted. Ideally, altimeters better suited for ice sheet elevation studies should be employed. The radar altimeters used thus far have been primarily designed for ocean wave height determination. But the functional response of the radar altimeter over continental ice sheets is considerably more complex than over the ocean surface (Martin *et al.*, 1983). In addition, a true polar-

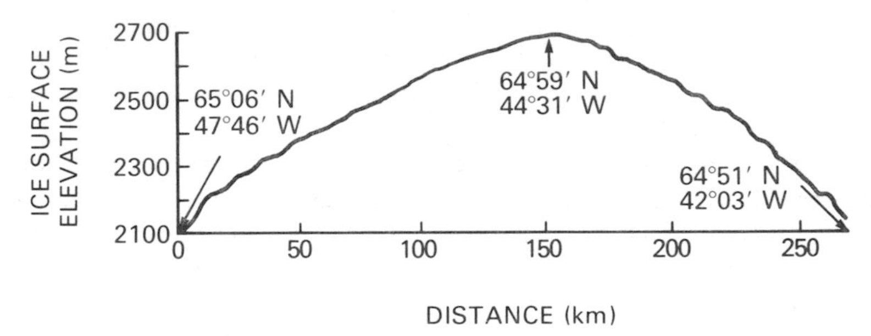

Fig. 7.16 Greenland Ice Sheet profile measured by the NOAA GOES-3 radar altimeter. Vertical exaggeration is 134/1 (from Brooks *et al.*, 1978). Reprinted by permission from *Nature* (London), **274**, 539–43. Copyright © 1978 Macmillan Journals Ltd.

orbiting spacecraft would optimize coverage of the continental ice sheets (Brooks *et al.*, 1978).

7.5 Icebergs

Because icebergs calve from glaciers, they are mentioned in this chapter. However, once calved, icebergs can travel through the sea ice and so they are mentioned again in Chapter 8 in the section on sea ice movement (Section 8.5). Icebergs can be a major hindrance to shipping and thus tracking of icebergs near shipping lanes is important. Icebergs exist in various sizes, from small ice fragments to tabular icebergs larger in area than the State of Rhode Island in the United States. Since icebergs calve from glaciers, they are composed of fresh water and may thus be potential sources of fresh water for human consumption if they can be transported to arid areas. The subject of Antarctic iceberg towing has received some attention in the last few years. If large icebergs that break off from Antarctic glaciers and ice shelves could be towed through the ocean without considerable melting, then a continued supply of fresh water could possibly be available to areas that have fresh water shortages. Towing distance, ocean currents and iceberg deterioration rate are major factors to consider in the feasibility of this technique (Donaldson, 1978). Satellite data may prove useful for locating icebergs suitable for towing. Towing of Antarctic icebergs to Australia or to the west coast of South America or Africa appears far more feasible than does towing to locations in the Northern Hemisphere (Holden, 1977).

Another consideration is the detection of icebergs in shipping routes. An iceberg represents a very large volume of ice above and below the surface of the ocean and, as such, can be extremely hazardous to ships. Also, icebergs imbedded in sea ice can look very similar to sea ice. The ability to distinguish between icebergs and sea ice can be quite important. (See Chapter 8, Section 8.5 for additional information on this subject.) In addition, icebergs in or near shipping routes may need to be tracked regardless of weather (or cloud cover) conditions. In fact, on the Grand Banks in the Labrador Sea, the maximum occurrence of icebergs coincides with the maximum occurrence of fog (Larson *et al.*, 1978). Thus, a need for the use of microwave remote sensing for iceberg tracking is indicated.

Both passive and active microwave remote sensing has been used to identify and track icebergs. Passive microwave remote sensing does not have a resolution suitable for studies of small icebergs, but has been used to track large icebergs even from satellites. For example, in the Weddell Sea near Antarctica, a series of images from the Nimbus 5 ESMR satellite was used to track large icebergs that had calved from Antarctic ice shelves. A tabular iceberg has an emissivity that is lower than first year sea ice and thus is discernable from the surrounding sea ice using T_B data (Zwally and Gloersen, 1977).

Smaller icebergs can often be detected using SLR (Side Looking Radar) images. The many facets of an iceberg reflect the radar signal causing them to be discernable from surrounding (smoother) sea ice using radar data. However, detection of icebergs can be greatly complicated when they are surrounded by ice clutter such as multiyear sea ice. Small ice floes that have frozen together and have a lot of ridges will cause high radar backscatter similar to that produced by icebergs (Larson *et al.*, 1978). Nevertheless, Larson *et al.* (1978) were able to detect icebergs in a variety of sea ice clutter, to identify icebergs by shape characteristics and to estimate iceberg size by radar shadow information.

Kirby and Lowry (1981) found that the shorter wavelength X-band SAR sensor was better for iceberg detection than the L-band SAR sensor. They studied SAR data along the Greenland coast from Disko Island to Thule. Using the X-band imagery, icebergs were classified according to size and type, but this was not possible with L-band imagery. L-band penetration into freshwater ice is greater than X-band penetration and thus iceberg surface detail is greatly reduced using the L-band data, rendering it similar to the surrounding sea ice clutter. The orientation of the iceberg with respect to the radar is another factor affecting the detectability of the icebergs (Kirby and Lowry, 1981).

Landsat data have been investigated for use in iceberg studies but their utility is severely limited by cloudcover and darkness. Hult and Ostrander (1974) used Landsat imagery to locate clusters of icebergs near Antarctica, but found that the areas of interest were cloudcovered so often that iceberg tracking and studies of seasonal behavior of icebergs were precluded.

7.6 Radio echo sounding of glacier ice

Probably the most difficult part of analysis using radio echo sounding (RES) is the correct interpretation of the echos produced from internal layering. Numerous theories have been advanced regarding the significance of these layers. When radio echo sounding techniques were first used on glacier ice, they were developed to measure ice thickness, and scattering from internal layers was considered 'clutter'. However, RES of glacier ice has evolved to the point that internal scattering is considered to provide useful information. This has furthered our understanding of ice sheet and valley glacier structure and dynamics. When sounding polar ice from aircraft, separate echos from upper and lower ice sheet surfaces are obtained. Thus determination of ice thickness is possible. The signal from the upper surface is normally very strong and the strength of the lower echo is weaker and nearly continuous but varies rapidly (Robin, 1975).

Internal reflections are caused by changes in the electrical permittivity within the ice (inhomogeneities) or by dielectric changes between layers. Propagation of radio waves through ice masses is controlled by their refraction, absorption and reflection. The formula:

$$V = c/\epsilon^{1/2} \tag{7.3}$$

where V is phase velocity, c is the velocity of electromagnetic waves in space and ϵ is the permittivity of polycrystalline ice, describes the velocity of the electromagnetic wave through ice. From laboratory studies $V = 169\,\mathrm{m}/\mu\mathrm{s}$ when mean values of ϵ are used (Robin, 1975). As ice density increases, radar wave velocities decrease.

Inhomogeneities that cause internal reflections consist of ice lenses and layers from meltwater percolation, rock material and crevasses (Macheret and Zhuravlev, 1982), presence of impurities (e.g. ash layers or chemical precipitates), fluctuations in ice density, presence of brine or changes in temperature with depth (Dowdeswell *et al.*, 1984). Ice in temperate glaciers presents more difficulties for RES of ice thickness than does ice in the two polar ice sheets because there are more internal scatterers (inhomogeneities) in temperate glaciers. (The Greenland Ice Sheet, however, has characteristics of a temperate glacier in the southern part while the northern part is a true polar glacier.)

Large areas (several hundred km^2) of reflective 'layers' have been found in both the Greenland and Antarctic ice sheets (Gudmandsen, 1975 and Drewry, 1981). In Greenland, Gudmandsen (1975) reported that the age of the ice in the deepest observable layer coincided with the timing of the last glaciation and correlated to some degree with the $\delta(O^{18})$ record. (Two of the stable isotopic forms of oxygen, O^{16} and O^{18}, depend mainly on the temperature of condensation at the time of ice formation (Flint, 1971). The ratio O^{18}/O^{16} is expressed as the deviation $\delta(O^{18})$ in parts per thousand from a standard. High values of $\delta(O^{18})$ are indicative of the warmest climate.) Gudmandsen (1975 and 1977) reported a multitude of echo-producing layers in the upper 2000 m of the ice in Greenland. Because of the proximity of Greenland to Iceland (an active volcanic region), some of these subglacial reflecting layers may be deposits of volcanic ash or tephra (R. Williams, written communication, 1984).

Airborne RES of 10^6 km^2 of West Antarctica have been obtained by a joint project involving the Scott Polar Research Institute, National Science Foundation and the Technical University of Denmark (Jankowski and Drewry, 1981). Pulsed radars at a center frequency of 60 and 300 MHz were used to conduct the first detailed, systematic study of bedrock in West Antarctica using RES. The ability to do the RES from the air permitted a detailed bedrock map to be made and thus a detailed study of bedrock to be conducted. Such work would not have been feasible if it had to be undertaken using ground-based soundings. Many important findings regarding the bedrock configuration resulted from this work. For example, a sinuous ridge standing 1000 m above the floor of the basin was found to extend across the center of the extensive Byrd Subglacial Basin. A well-defined boundary was found between the Byrd Subglacial Basin and a highland area extending southward from the Ellsworth Mountains.

Radio echo soundings of the ice sheet permitted the delineation of many of the ice drainage basins of West Antarctica. West Antarctica is divided into a series of drainage basins feeding into the Ross Ice Shelf, Ronne-Filchner Ice Shelf and the Amundsen-Bellingshausen Seas (Jankowski and Drewry, 1981). Surface crevassing as detected on the RES data revealed the ice streams.

Bottom crevasses of the Ross Ice Shelf were detected and mapped by Jezek and Bentley (1983) using surface and airborne RES. They found two major concentrations of crevasses and concluded that most of the crevasse sites are associated with ice rises. An ice rise is a portion of an ice shelf that is grounded on bedrock causing a dome-shaped surface on the ice shelf. A radial flow pattern extends outward from an ice rise and its marginal areas are extensively crevassed due to shear as the shelf ice flows around the ice rise (Paterson, 1981).

A 60 MHz Mark 4 sounder, mounted on a DHC-6 Twin Otter aircraft was used to compare the bedrock RES echos with ice surface elevations along flight tracks, thereby enabling the inland boundary of the Ronne Ice Shelf to be located (Swithinbank *et al.*, 1976). The flights took place in January of 1975. Surface elevations were established by continuous recording of the pressure altitude of the aircraft in relation to terrain clearance measured by radar. Comparison of the character of the bottom echo with surface elevations enabled Swithinbank *et al.* (1976) to locate the inland boundary of the Ronne Ice Shelf. Measurements obtained during the 3 days of flights were used to redefine the boundaries of the Ronne Ice Shelf. As a result, the previously-mapped surface area of the ice shelf was reduced by 11% while the size of the inland ice sheet was increased because of the change in the location of the boundary.

In a polar ice sheet, snow density increases gradually from the surface into solid ice. In addition, the ice becomes warmer and ice density fluctuations decrease with depth (Paren and Robin, 1975). From work in central Antarctica, Paren and Robin (1975) found that ice density fluctuations account for echos at depths above 1500 m, but echos attributed to deeper layers were better explained by variations in ice conductivity. Clough (1977) also noted that ice density variations are the most likely cause of internal layering observed in the top 1000 m of the Greenland and Antarctic ice sheets. Elachi and Brown (1975) recorded up to 12 layers in the upper 100 m of the Greenland Ice Sheet using a synthetic aperture radar (SAR) sounder operating at a frequency of 150 MHz. Reflections from some internal layers were so strong that the strength of the return was greater than the return from the surface.

In Antarctica, strong bottom reflections have been detected using radio-echo sounding. These highly reflective areas, found to occupy bedrock hollows, have been interpreted as areas of sub-ice lakes filled with liquid water (Drewry, 1981). The largest yet discovered are approximately 8000 km^2, though most are much smaller (Drewry, 1981). These areas of the ice sheet are at the pressure melting point. Oswald (1975) confirms the existence of lakes of liquid water beneath the ice of East Antarctica and suggests that basal melting occurs over wide areas in East Antarctica.

Landsat imagery was used as a base 'map' for displaying the results of airborne RES of the Lambert Glacier basin in East Antarctica (Morgan and Budd, 1975) and for plotting surface elevation contours (1 and 5 m) on the Amery Ice Shelf (Brooks *et al.*, 1983). The Lambert Glacier and Amery Ice Shelf system drains a large part of the East Antarctic Ice Sheet. This drainage is accomplished through a relatively narrow outlet. Knowledge of the volume of ice drainage is important for analysis of the mass balance of the Antarctic Ice Sheet. Measurement of both ice velocity and ice thickness are required. The Australian National Antarctic Research Expeditions have undertaken an extensive program to study the Amery Ice Shelf and the Lambert Glacier basin. A standard radar altimeter operating at 440 MHz, and a barometric altimeter were used for surface elevation determination, and a 100 MHz radio echo sounder was employed for ice thickness measurements. Maximum ice thickness was measured to be approximately 2500 m in the center of the Lambert Glacier. Flight lines, surface elevation, and ice thickness were drawn onto MSS band 7 Landsat imagery. In the lower reaches of the Lambert Glacier, strong echos indicated the presence of considerable basal meltwater (Morgan and Budd, 1975). Figure 7.17 shows vertical section profiles obtained from the soundings.

Three cold valley glaciers in the Steele Creek drainage basin, St Elias Mountains, Yukon Territory, Canada were studied using airborne RES at a center frequency of 840 MHz, in the UHF (ultra high frequency) band (Narod and Clarke, 1980). On the Rusty and Trapridge Glaciers, measurements of ice thickness using helicopter-borne RES were found to compare well with measurements obtained by ground-based sounders. The third glacier, Hazard Glacier, apparently has an internal moraine layer formed by the merging of two main tributary glaciers, thus causing some difficulty in obtaining accurate ice thicknesses on that glacier. Goodman *et al.* (1975) used surface RES at a frequency of 620 MHz to measure the thickness of ice on the Trapridge Glacier, a surge-type glacier, and found a maximum thickness of 143 m. A rapid change was detected in the lower ablation region of the glacier, marking the boundary between active and stagnant ice. This was related to basal water conditions which might influence the surge behavior of the glacier (Goodman *et al.*, 1975).

RES data of temperate glaciers have more echo-producing horizons than do such data of ice sheets. The numerous layers present are caused by internal reflections from subsurface inhomogeneities as mentioned previously. Macheret and Zhuravlev (1982) studied 87 valley glaciers in Svalbard using RES at a center frequency of 620 MHz. They found that the best time to study the temperate glaciers was from March to May (before the beginning of the melt season). They flew 2000 km of airborne RES profiles. To corroborate the measurements, repeated flights using the RES were made along the same flight lines and ground-based RES measurements were made as well. Internal

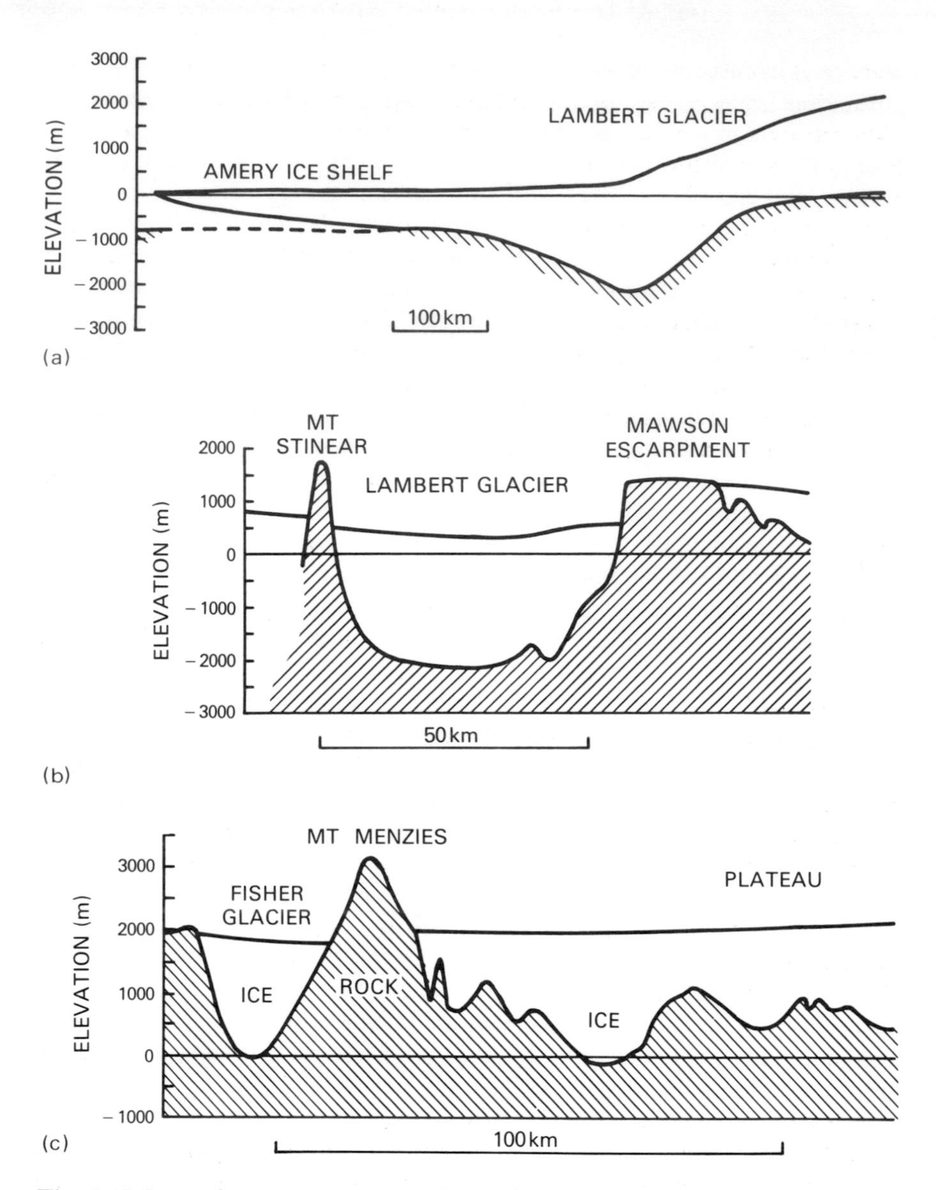

Fig. 7.17 Ice surface and bedrock profiles obtained using radio echo sounding (from Morgan and Budd, 1975). (a) Ice surface and bedrock profile along a flow line from the inland ice sheet through the Lambert Glacier to the Amery Ice Shelf and the coast. Smooth profiles were drawn from the data obtained at 3-km intervals. (b) Ice surface and bedrock profiles across the deep section of the Lambert Glacier near Mt Stinear. (c) Ice surface and bedrock profile across Mt Menzies, the Fisher Glacier to the north, and the plateau to the south. (Reprinted by permission of the International Glaciological Society from *J. Glaciol.*, **15**, 103–11.)

reflecting boundaries were located at approximately 70–180 m depth in some Svalbard glaciers that are 250–300 m thick. Such a boundary may have resulted from: (1) differences in water content of the ice, (2) presence of water or morainic material within the ice or (3) layers formed in the past due to climatic warming (Macheret and Zhuravlev, 1982).

Recent work by Dowdeswell *et al.* (1984) indicates that Macheret and Zhuravlev (1982) underestimated the ice thickness of the Svalbard glaciers by two to three times. Using a 60 MHz sounder, Dowdeswell *et al.* (1984) found internal reflecting horizons on some Svalbard glaciers that they believe corresponded to Macheret and Zhuravlev's presumed bottom echos. The lower frequency radio echo sounding equipment (60 MHz) used by Dowdeswell *et al.* (1984) was found to allow increased penetration into the temperate glaciers as compared to the higher frequency equipment (620 MHz) used by Macheret and Zhuravlev (1982).

Radio echo sounding of ice sheets and temperate valley glaciers has become a powerful tool for the study of glacier ice. Combined with data from other airborne and ground-based sensors, RES can give an excellent picture of the thickness, flow patterns, internal layering and inhomogeneities of glaciers. Unraveling the precise reasons for the formation of such layers is proving to be a complex but fascinating and revealing aspect of glaciology.

References

Benson, C.S. (1962) *Stratigraphic Studies in the Snow and Firn of the Greenland Ice Sheet*, US Army Cold Regions Research and Engineering Laboratory, Hanover, NH, CRREL Research Report No. 70.

Bindschadler, R.A. (1984) Jakobshavns Glacier Drainage Basin: A balance assessment. *J. Geophys. Res.*, **89**, 2066–72.

Bjornsson, H. (1980) The surface area of glaciers in Iceland. *Jökull*, **28**, 31.

Brenner, A.C., Bindschadler, R.A., Thomas, R.H. and Zwally, H.J. (1983) Slope-induced errors in radar altimetry over continental ice sheets. *J. Geophys. Res.*, **88**, 1617–23.

Brooks, R.L., Campbell, W.J. and Ramseier, R.O. *et al.*, (1978) Ice sheet topography by satellite altimetry. *Nature (London)*, **274**, 539–43.

Brooks, R.L., Williams, R.S. Jr, Ferrigno, J.G. and Krabill, W.B. (1983) Amery ice shelf topography from satellite radar altimetry. *In Antarctic Earth Science* (eds R.L. Oliver, P.R. James and J.B. Jago), Proceedings of the Fourth International Symposium on Antarctic Earth Sciences, 16–20 August 1984, University of Adelaide, South Australia, Australian Academy of Sciences, Canberra, pp. 441–5.

Cassidy, W.A., Muenier, T., Buchwald, V. and Thompson, C., (1984) Search for meteorites in the Allan Hills/Elephant Moraine area, 1982–1983. *Antarct. J. US*, **18**, 81–2.

Chang, A.T.C., Gloersen, P. and Schmugge, T. *et al.* (1976) Microwave emission from snow and glacier ice. *J. Glaciol.*, **16**, 23–39.

Clough, J.W. (1977) Radio-echo sounding: reflections from internal layers in ice sheets.

J. Glaciol. **18**, 3–14.

Comiso, J.C., Zwally, H.J. and Saba, J.L. (1982) Radiative transfer modeling of microwave emission and dependence on firn properties. *Ann. Glaciol.*, **3**, 54–8.

Crabtree, R.D. (1976) Changes in the Mýrdalsjökull ice cap, south Iceland: possible uses of satellite imagery. *Polar Rec.*, **18**, 73–6.

Donaldson, P.B. (1978) Melting of Antarctic icebergs. *Nature (London)*, **275**, 305–6.

Dowdeswell, J.A., Drewry, D.J., Liestøl, O. and Orheim, O. (1984) Radio echo-sounding of Spitsbergen Glaciers: problems in the interpretation of layer and bottom returns. *J. Glaciol.*, **30**, 16–21.

Drewry, D.J. (1981) Radio echo sounding of ice masses: principles and applications. In *Remote Sensing in Meteorology, Oceanography and Hydrology* (ed. A.P. Cracknell). Ellis Horwood Ltd, Chichester, pp. 270–84.

Elachi, C. and Brown, W.E. Jr, (1975) Imaging and sounding of ice fields with airborne coherent radars. *J. Geophys. Res.*, **80**, 1113–19.

Ferrigno, J.G. and Williams, R.S. Jr, (1983) Limitations in the use of Landsat images for mapping and other purposes in snow- and ice-covered regions: Antarctica, Iceland, and Cape Cod, Massachusetts, *Proceedings of the Seventeenth International Symposium on Remote Sensing of Environment*, Vol. 1, Environmental Research Institute of Michigan, Ann Arbor, MI, pp. 335–55.

Field, W.O. (ed.) (1975) *Mountain Glaciers of the Northern Hemisphere*, Vol. 2, US Army Cold Regions Research and Engineering Laboratory, CRREL, Hanover, NH, pp. 299–492.

Flint, R.F. (1971) *Glacial and Quaternary Geology*. John Wiley, New York.

Goodman, R.H., Clarke, G.K.C. and Jarvis, G.T. *et al.* (1975) Radio soundings on Trapridge Glacier, Yukon Territory, Canada. *J. Glaciol.*, **14**, 79–84.

Gudmandsen, P. (1975) Layer echoes in polar ice sheets. *J. Glaciol.*, **15**, 95–101.

Gudmandsen, P. (1977) Studies of ice by means of radio echo sounding. In *Remote Sensing of the Terrestrial Environment* (eds R.F. Peel, L.F. Curtis and E.C. Barrett), Butterworths, London, pp. 198–211.

Hall, D.K. and Ormsby, J.P. (1983) Use of SEASAT synthetic aperture radar and LANDSAT multispectral scanner subsystem data for Alaskan glaciology studies. *J. Geophys. Res.*, **88**, 1597–607.

Holden, C. (1977) Experts ponder icebergs as relief for world water dilemma. *Science*, **198**, 274–6.

Hult, J.L. and Ostrander, N.C. (1974) Applicability of ERTS to Antarctic Iceberg Resources, *Third Earth Resources Technology Satellite-1 Symposium*, Vol. 1, Technical presentations, Section B, NASA SP-351, National Aeronautics and Space Administration, Washington, DC, pp. 1467–90.

Jankowski, E.J. and Drewry, D.J. (1981) The structure of West Antarctica from geophysical studies. *Nature (London)*, **291**, 17–21.

Jezek, K.C. and Bentley, C.R. (1983) Field studies of bottom crevasses in the Ross Ice Shelf, Antarctica. *J. Glaciol.*, **29**, 118–26.

Kirby, M.E. and Lowry, R.T. (1981) Iceberg detectability problems using SAR and SLAR systems. In *Satellite Hydrology* (eds M. Deutsch, D.R. Wiesnet and A. Rango), Proceedings of the Fifth Annual William T. Pecora Memorial Symposium on Remote Sensing, 10–15 June 1979, American Water Resources Association, Minneapolis, MN, pp. 200–12.

Krimmel, R.M. and Meier, M.F. (1975) Glacier applications of ERTS images. *J. Glaciol.*, **15**, 391–402.

Krimmel, R.M., Post, A. and Meier, M.F. (1976) Surging and nonsurging glaciers in the Pamir Mountains, USSR. In *ERTS-1, A New Window on our Planet* (eds R.S. Williams, Jr and W.D. Carter), US Geological Survey Professional Paper 929, pp. 178–84.

Larson, R.W., Shuchman, R.A. Rawson, R.A. and Worsfold, R.D. (1978) The use of SAR systems for iceberg detection and characterization, *Proceedings of the Twelfth International Symposium on Remote Sensing of Environment*, Vol. 2, Environmental Research Institute of Michigan, Ann Arbor, MI, pp. 1127–47.

Macheret, Yu. Ya. and Zhuravlev, A.B. (1982) Radio echo-sounding of Svalbard glaciers. *J. Glaciol.*, **28**, 295–314.

Martin, T.V., Zwally, H.J., Brenner, A.C. and Bindschadler, R.A. (1983) Analysis and retracking of continental ice sheet radar altimeter waveforms. *J. Geophys. Res.*, **88**, 1608–16.

Mayo, L.R. (1978) Identification of unstable glaciers intermediate between normal and surging glaciers, *Proceedings of the International Workshop on Mechanism of Glacier Variations*, Academy of Sciences of the USSR, Section of Glaciology of the Soviet Geophysical Committee and Institute of Geography, Moscow, pp. 133–5.

Meier, M.F. (1973) Evaluation of ERTS imagery for mapping and detection of changes in snowcover on land and on glaciers, *Symposium on Significant Results Obtained from the Earth Resources Technology Satellite-1*, National Aeronautics and Space Administration, Washington DC, Vol. 1, pp. 863–75.

Meier, M.F. (1976) Monitoring the motion of surging glaciers in the Mount McKinley Massif, Alaska. In *ERTS-1, A New Window on our Planet* (eds R.S. Williams, Jr and W.D. Carter), US Geological Survey Professional Paper 929, pp. 185–7.

Meier, M.F., Rasmussen, L.A., Post, A. *et al.* (1980) *Predicted Timing of the Disintegration of the Lower Reach of Columbia Glacier, Alaska*, US Geological Survey Open File Report 80–582.

Morgan, V.I. and Budd, W.F. (1975) Radio-echo sounding of the Lambert Glacier basin. *J. Glaciol.*, **15**, 103–11.

Narod, B.B. and Clarke, G.K.C. (1980) Airborne UHF radio echo sounding of three Yukon Glaciers. *J. Glaciol.*, **25**, 23–31.

Østrem, G. (1975) ERTS data in glaciology – an effort to monitor glacier mass balance from satellite imagery. *J. Glaciol.*, **15**, 403–15.

Oswald, G.K.A. (1975) Investigation of sub-ice bedrock characteristics by radio-echo sounding. *J. Glaciol.*, **15**, 75–87.

Paren, J.G. and Robin, G. de Q. (1975) Internal reflections in polar ice sheets. *J. Glaciol.*, **14**, 251–9.

Paterson, W.S.B. (1981) *The Physics of Glaciers*, 2nd edn, Pergamon Press, New York.

Post, A. (1960) The exceptional advances of the Muldrow, Black Rapids, and Susitna Glaciers. *J. Geophys. Res.*, **65**, 3703–12.

Post, A., Meier, M.F. and Mayo, L.R. (1976) Measuring the motion of the Lowell and Tweedsmuir surging glaciers of British Columbia, Canada. In *ERTS-1, A New Window on Our Planet* (eds R.S. Williams, Jr and W.D. Carter), US Geological Survey Professional Paper 929, pp. 180–4.

Robin, G. de Q. (1975) Radio-echo sounding: glaciological interpretations and applications. *J. Glaciol.*, **15**, 49–64.

Sikonia, W.G. and Post, A. (1979) *Columbia Glacier, Alaska – Recent Ice Loss and its Relationship to Seasonal Terminal Embayments, Thinning and Glacier Flow*, US Geological Survey Open File Report 79-1265.

Sugden, D.E. and John, B.S. (1976) *Glaciers and Landscape*, John Wiley, New York.

Swithinbank, C., Doake, C., Wager, A. and Crabtree, R. (1976) Major changes in the map of Antarctica. *Polar Rec.*, **18**, 295–9.

Thorarinsson, S., Saemundsson, K. and Williams, R.S. Jr (1973) ERTS-1 Image of Vatnajökull: Analysis of glaciological, structural, and volcanic features. *Jökull*, **23**, 7–17.

Tomasson, H. (1975) Grímsvatnahlaup 1972, mechanism and sediment discharge. *Jökull*, **24**, 27–39.

Williams, R.S., Jr (1976) Vatnajökull Icecap, Iceland. In *ERTS-1, A New Window on our Planet* (eds R.S. Williams, Jr and W.D. Carter), US Geological Survey Professional Paper 929, pp. 188–93.

Williams, R.S., Jr (1983a) Satellite Glaciology of Iceland. *Jökull*, **33**, 3–12.

Williams, R.S., Jr (ed.) (1983b) Geological applications. *Manual of Remote Sensing*, 2nd edn, American Society of Photogrammetry, Falls Church, VA, Ch. 31, pp. 1667–953.

Williams, R.S., Jr, Bodvarsson, A. and Fridrksson, S. *et al.* (1974) Environmental studies of Iceland with ERTS-1 imagery, *Proceedings of the Ninth Symposium on Remote Sensing of Environment*, Vol. 1, Environmental Research Institute of Michigan, Ann Arbor, MI, pp. 31–81.

Williams, R.S., Jr and Ferrigno, J.G. (1981) Satellite image atlas of the Earth's glaciers. In *Satellite Hydrology* (eds M. Deutch, D.R. Wiesnet and A. Rango), Proceedings of the Fifth Annual William T. Pecora Memorial Symposium on Remote Sensing, 10–15 June, 1979, American Water Resources Association, MN, pp. 173–82.

Williams, R.S., Jr, Ferrigno, J.G. and Kent, T.M. (1984a) *Index Map Table showing Optimum Landsat 1, 2 and 3 Images of Antarctica*, US Geological Survey Open-File Report 84–573.

Williams, R.S., Jr, Ferrigno, J.G. and Meunier, T.K. (1984b) Satellite glaciology project. *Antarct. J. US*, **18**, 119–21.

Williams, R.S., Jr, Meunier, T.K. and Ferrigno, J.G. (1983) Blue ice, meteorites and satellite imagery in Antarctica. *Polar Rec.*, **21**, 493–6.

Williams, R.S., Jr, Thorarinsson, S., Bjornsson, H. *et al* (1979). Dynamics of Icelandic ice caps and glaciers. *J. Glaciol.*, **24**, 505–7.

Zwally, H.J. (1977) Microwave emissivity and accumulation rate of polar firn. *J. Glaciol.*, **18**, 195–215.

Zwally, H.J., Bindschadler, R.A. and Brenner, A.C. *et al.* (1983) Surface elevation contours of Greenland and Antarctic ice sheets. *J. Geophys. Res.*, **88**, 1589–96.

Zwally, H.J. and P. Gloersen (1977) Passive microwave images of the polar regions and research applications. *Polar Rec.*, **18**, 431–50.

8

Sea ice

8.1 Introduction

Sea ice (Figs 8.1, 8.2 and 8.3) is present over approximately 13% of the Earth's ocean surface (Weeks, 1981). It is a highly variable feature and its presence or absence at any given time has a profound effect on the Earth's radiation budget. The albedo of ice-covered ocean is dramatically higher than that of open water. Additionally, the ice cover is an insulating layer between the ocean and atmosphere; heat loss through open water is approximately 100 times greater than heat loss through thick ice. As a consequence, leads and polynyas (linear and non-linear openings in sea ice) are significant to the energy budget of the ice-covered ocean and to local and regional climatology. Such open water areas and areas of reduced ice concentration are also important for shipping in ice-covered seas.

As it ages, newly formed, smooth and thin sea ice is metamorphosed by temperature fluctuations, compressive and shear forces, surface currents and winds. In addition, the ice thickens and snow falls on top of the ice. Ridge formation and surface roughness increase with age, and the angular edges and smooth surfaces of first-year ice floes (Fig. 8.3) are transformed into rounded edges with hummocky, ridged surfaces.

Another important process that occurs as sea ice ages is desalinization. Sea water contains 30–35 parts per thousand salt. Brine drains out or is flushed out of the ice as the ice freezes and ages. If sea ice could be frozen extremely slowly, pure ice would result because of rejection of all salts during the freezing process. However, the freezing rate is normally too rapid to allow pure ice to form, thus brine is trapped in the lattice structure of the ice (Pounder, 1965). Both meltwater percolation and gravity expedite brine drainage. Typical sea ice

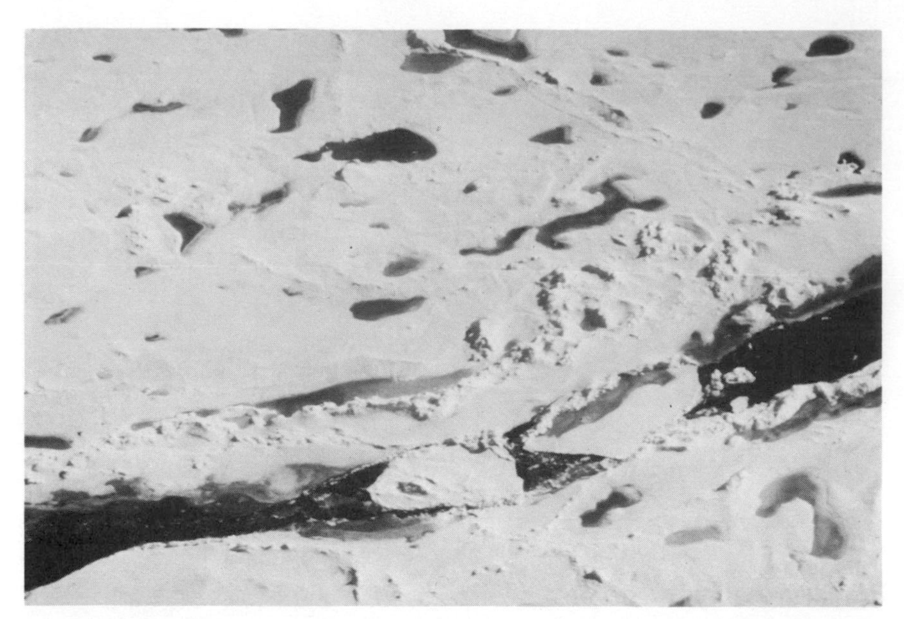

Fig. 8.1 Photograph of multiyear sea ice in the Beaufort Sea – August 1975 (photograph by C. Parkinson).

Fig. 8.2 Grounded sea ice near Tapkaluk Island southwest of Pt Barrow, AK – October 1977 (photograph by I. Virsnieks).

Fig. 8.3 First-year sea ice in MIZEX-West – February 1983 (photograph by I. Virsnieks).

salinities are 10–15 parts per thousand (‰) for newly formed sea ice, 5–8‰ for 1–2 m thick first-year ice and 0.1–3.0‰ for multiyear sea ice (Weeks, 1981).

The brine drainage process results in numerous gas bubbles within the ice. These spherical bubbles are effective scatterers of microwave emission and contribute to a lower microwave T_B in multiyear as compared to first-year sea ice. Thus passive microwave data are useful for distinguishing between first-year and thicker multiyear sea ice.

Radar data have been used for distinguishing between first-year and multiyear sea ice because radars sense surface roughness, and the roughness often increases as the sea ice ages. Higher backscatter from the rougher surface of the multiyear floes, relative to first-year floes, can result especially if different ice types are observable on the same image. However, the use of side looking radar to unambiguously distinguish ice type is not recommended at this time.

Side looking radar (SLR) systems have been used extensively to study sea ice conditions both from aircraft and satellite platforms. It is feasible to use SLR to (Campbell *et al.*, 1975):

(1) Distinguish open or newly frozen leads from older ice
(2) Generally distinguish thin ice from open water
(3) Delineate pressure ridges and highly deformed ice
(4) Observe and measure ice floe shape and size
(5) Distinguish land from shore fast ice.

Table 8.1 Remote sensing platforms and sensors for AIDJEX (from Campbell *et al.*, 1978)

Sensor/sensor platform	*CV 990, NASA*	*C-47, CCRS**	*Argus, DND†*	*Flextrack*
Radiometer	Aerojet	Aerojet		Aerojet
frequency	19 GHz	37 GHz		4.99, 13.4, 37 GHz
polarization	1.8 cm	0.81 cm		6.0, 2.23, 0.81 cm
incident angle	nadir, ± 55° starboard and port scanning	45° forward		H, V all three, nadir to 55° forward
resolution	500 × 500 m	15 × 15 m		1.5 × 1.5, 0.7 × 0.7, 0.4 × 0.4 m
Imaging radar	JPL		Motorola AN/ADS 94D	
type	synthetic aperture		real aperture	
frequency	1.215 GHz		9.2 GHz	
wavelength	24.5 cm		3.25 cm	
polarization	HH		HH	
incident angle	0–55° starboard		45–88° starboard and port	
resolution				
range	25 m		30 m	
azimuthal	25 m		40–200 m	
swath	14 km		25 km	

Scatterometer		Ryan 720		
frequency		13.3 GHz		
wavelength		2.25 cm		
polarization		HH, HV, VV, VH		
incident angle		±60° fore and aft		
resolution		15 × 15 m		
Infrared		Daedalus		PRT-5
wavelength		8–14 μm		8–14 μm
incident angle		±55° starboard and port		nadir to 55° forward
Mapping camera	RC-10	RC-10		Handheld Nikon
Altitude data collected	12000 m	300 m	900 m	0 m (surface)

* Canadian Centre for Remote Sensing.
† Department of National Defense (Canada).

Visible and near-infrared NOAA VHRR and AVHRR data, Landsat MSS data and SLR data are useful for measuring and monitoring the movement of individual ice floes and for studying lead and polynya patterns. Thermal infrared data such as HCMM and NOAA data can be acquired during the long polar night, and both passive and active (radar) microwave data are obtainable through cloudcover and darkness. Landsat MSS data and SLR data have good resolution, but SLR data are often advantageous because data can be acquired irrespective of darkness and weather conditions and are therefore more versatile for studying ice movement and lead patterns.

Monitoring sea ice type and movement is of great importance in many northern countries. For example, in Canada most of the navigable waters, except along the Pacific coast and a portion of the Atlantic coast, are affected by floating ice during portions of the year. Extensive areas of the Canadian high Arctic are ice-covered all year. Ice and its management are central to the economic well-being of Canada (Hengeveld, 1974). Because of this need for sea ice information, the Canadians will launch a satellite with a C-band (5.3 GHz) HH polarized SAR on-board that is specifically designed to give near real-time information on sea ice conditions. 'Radarsat,' to be launched in 1990, will enable captains of oil tankers to select routes to the Northwest Passage that will allow them to travel the most efficient and safest routes possible (Langham, 1982).

Several international sea ice experiments have been conducted since the early 1970s. Experiments such as BESEX (Bering Sea Experiment), AIDJEX (Arctic Ice Dynamics Joint Experiment), NORSEX (Norwegian Remote Sensing Experiment), and MIZEX (Marginal Ice Zone Experiment) have employed a wide variety of remote sensors and platforms as well as intensive ground measurements to study sea ice. Results from these and other experiments have greatly contributed to our understanding of the dynamics, morphology and internal characteristics of sea ice, and the interaction of sea ice with ocean currents and the atmosphere.

During the BESEX, AIDJEX and NORSEX experiments, active and passive microwave instruments were employed and measurement techniques refined for the study of sea ice. The BESEX and AIDJEX missions utilized aircraft flights during various seasons to measure ice concentration over very large areas at intervals of several days. During the AIDEX main experiment in 1975 and 1976, surface and aircraft measurements were obtained in the Beaufort Sea during all seasons using a variety of aircraft, satellite-borne and ground-based sensors (Campbell *et al.*, 1978) (Table 8.1).

NORSEX provided sequential imaging radar data from aircraft on the structure of the marginal ice zone, ocean eddies and waves. The intent of the on-going MIZEX mission is to gain an improved understanding of the processes occurring in the marginal ice zone. Two ice margin areas are being studied intensively: (1) a deep water ice edge in the Fram Strait region in terms of lateral

Table 8.2 Remote sensing program for MIZEX (after Wadhams *et al.*, 1981)

Platform	*Sensor*	*Purpose*	*Frequency of coverage*
Aircraft	SAR	map sea ice in experimental area	every 3 days
Aircraft	passive MW	study ocean and ice characteristics	every 3 days
Aircraft	IR and Photography	study ice surface characteristics	every 3 days
Landsats 4 and 5	MSS and TM	study ice surface characteristics	at least once every 16 days, cloud cover permitting
NOAA	AVHRR	study position and shape of ice edge	once or twice daily, cloud cover permitting
TIROS-N	AVHRR	study position and shape of ice edge	twice daily, cloud cover permitting
Aircraft	gust probe	measure wind stress	at least every 3 days

and vertical heat and momentum exchanges in the air-ice-surface system; and (2) a continental shelf ice edge area in the Bering Sea in which analysis of ice dynamics and deformation is being emphasized (Wadhams *et al.*, 1981). The remote sensing portion of MIZEX is summarized in Table 8.2.

There is no sensor presently operable that is ideal for the comprehensive study of sea ice. Combinations of sensors must be employed. The passive microwave satellite data are obtainable daily and through cloud cover and are useful for determination of ice type. However, the resolution is too poor for detailed studies of ice movement and lead structure. The imaging sensors on-board the Landsat and NOAA satellites are useful for ice movement and lead orientation studies, but all-too-frequently cloud cover intervenes to reduce the utility of these sensors. SLR is unsurpassed among remote sensors for showing lead orientation, shear zones and drift patterns throughout the year, day or night (Dunbar, 1975). In the future, it is expected that results from Canada's Radarsat and the international MIZEX experiment, will represent major contributions to the understanding of the properties of sea ice and the role of sea ice in climatic and oceanographic process.

8.2 Sea ice age

The determination of sea ice age, and particularly the distinction between first-year and multiyear ice is important for many reasons. Multiyear ice generally presents a greater hindrance for ship travel than does first-year ice because it is usually thicker and has more ridges. Study of first-year ice formation and movement can reveal important information about ice genesis and the mechanics of ice formation and transport. The role of ocean currents and winds can also be analyzed for specific areas. Aircraft and satellite studies using the horizontally polarized 1.55 cm Electrically Scanning Microwave Radiometer (ESMR) demonstrated that first-year sea ice has a higher emissivity at the 1.55 cm wavelength than does multiyear sea ice (Zwally and Gloersen, 1977). Specifically, first-year ice has an emissivity of about 0.92 and multiyear sea ice has an emissivity of approximately 0.84. Recall from Chapter 1 that emissivity and microwave T_B are related by: $\epsilon = T_B/T_s$ where ϵ is the emissivity, T_B is the brightness temperature as measured by a radiometer, and T_s is the surface temperature. The difference in ice emissivities leads to a difference in the recorded brightness temperatures, allowing one to use the T_Bs to distinguish relative age of ice and thereby to infer relative ice thickness.

Though analysis of brightness temperature data is useful for distinguishing between first-year and multiyear sea ice, there are important exceptions to this in the Arctic and Antarctic. In the Arctic, Cavalieri *et al.* (1984) show that the ability to distinguish first-year and multiyear ice using passive microwave remote sensing is not reliable beginning in July and lasting through September each year. This is due to the presence of surficial meltwater that obliterates the radiance from the ice below. In the Weddell Sea region of the southern ocean, multiyear sea ice is radiometrically similar to first-year ice (Zwally *et al.*, 1983a). This is because of low surface melt in the summers in this area thus impeding the brine drainage process. In addition, the high percentage (approximately 50%) of frazil ice in this area contributes to higher salinities because frazil ice effectively entraps and retains brine (Gow *et al.*, 1982). Based on amount of brine, the ice appears newer on the microwave data than it really is.

Ice age (first-year or multiyear) can be qualitatively determined in some cases using visible, infrared and radar data. It has long been recognized that multiyear ice floes have greater surface roughness due to greater deformation than first-year floes. The edges of multiyear floes are not as angular as are the edges of first-year floes. Visible, infrared and radar data are all useful for determining ice floe shape; and radar imagery is especially useful for detecting surface roughness. Different ice types as seen on Seasat SAR images are shown in Fig. 8.4.

Ambiguities in the interpretation of surface roughness can result especially if snow is present on the ice. This is particularly true if the snow is wet. Thus smooth, first-year ice can give high returns similar to those found for multiyear ice. Ketchum (1983) reported high, homogeneous X-band radar returns over snow-covered sea ice in Baffin Bay. The high moisture content of the snow caused such high returns. Ambiguities resulting from snow cover can usually be resolved if simultaneous X and L-band imagery is available. The L-band will generally penetrate thin snow overlying sea ice, whereas the X-band will not. High L-band returns are associated with a rough surface geometry and possibly sea ice subsurface effects (Ketchum, 1983). Additional ambiguities in interpretation of radar imagery of snow-covered sea ice occur in the marginal ice zone where there is a considerable likelihood of thin ice and large temperature fluctuations. These factors may contribute to higher X-band backscatter even from smooth first-year ice because of recrystallization of ice and snow in the surface and near-surface layers of the snow-covered sea ice. Again, L-band radar signals which are less scattered by moist snow should be employed to improve the interpretation especially in the marginal ice zone (Ketchum, 1983).

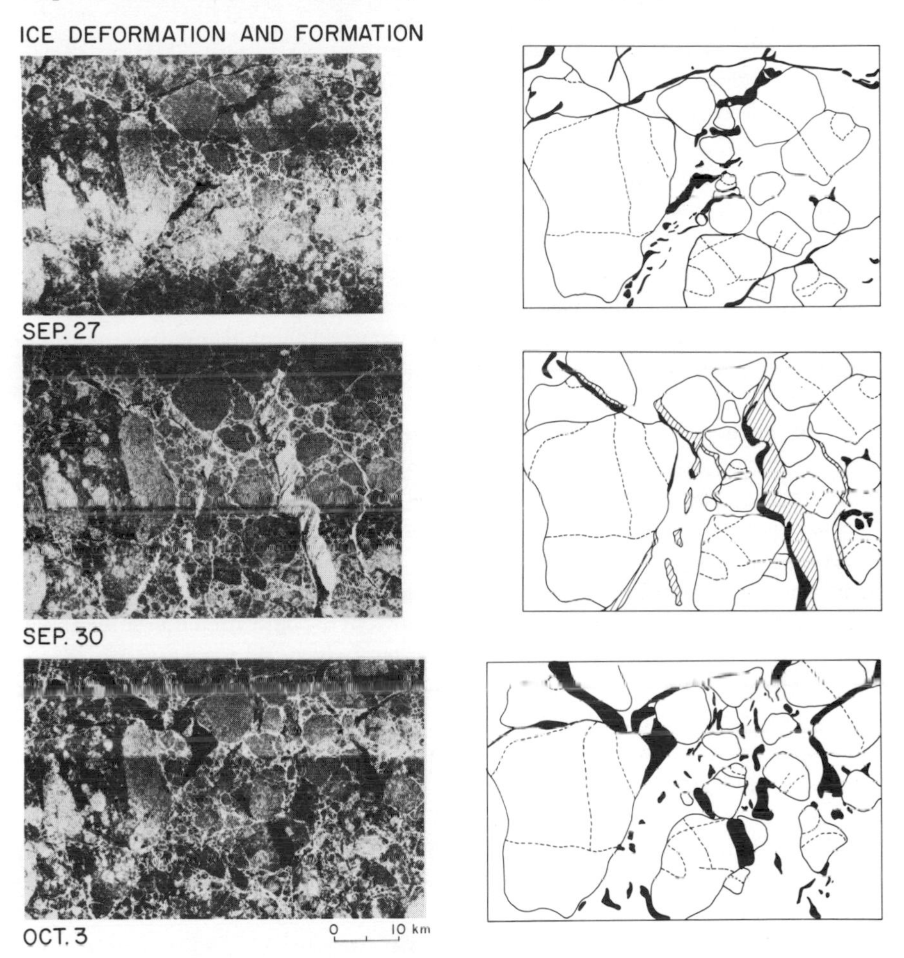

Fig. 8.4 Enlargements of three Seasat SAR images taken over a 6-day period in Melville Sound, Canada. The corresponding sketch maps indicate major flies (solid lines) with some distinguishing features (dashed lines), leads and/or new ice (dark zones), and grease ice (striped zones). Arrows indicate adjoining swaths: 27, September, 1978; 30 September, 1978 and 3 October, 1978 (from NMASA, 1983, courtesy of Jet Propulsion Laboratory).

8.3 Sea ice type and interannual variability

Passive microwave, radar, visible and thermal infrared satellite and aircraft imagery are all useful for interpreting sea ice types. However, satellite data are better adapted for analyzing the interannual variability of sea ice. While weather conditions may preclude aircraft reconnaissance flights for monitoring sea ice variability, passive and active microwave satellite data are acquired regardless of weather, and in the polar areas data quality is usually unaffected by cloud cover.

Bryan (1976) developed an interpretation key for SAR L-band HH polarized imagery of sea ice. He employed tone, linearity, texture and shape of imaged ice feature to identify and classify ice features in the Arctic Ocean. For example, a black, non-linear feature could be (1) first-year ice, (2) a melt pond or (3) a polynya; and a white linear feature could be (1) a ridge, (2) rafting or (3) the edge of a floe. Of course radar sea ice interpretation techniques must be keyed to geographic area, seasonal changes and radar parameters (Dunbar, 1975 and Ketchum, 1983). Under winter ice conditions, Lyden *et al.* (1984) found that X-band SAR data were more useful than L-band data for discriminating various ice types. Additionally, they found imagery obtained at small or steep incidence angles showed greater tonal variations between ice types than imagery obtained at larger angles.

Using Landsat imagery, ice floes with significant amounts of surface meltwater can be delineated by comparing MSS bands 5 and 7. Lower reflectance in band 7 (near-infrared) is indicative of the presence of surface meltwater. In Hendriksen Strait in the Queen Elizabeth Islands, Rango *et al.* (1973) found active melting between 23 and 28 August 1972 by comparing MSS bands 5 and 7. They noted a 50% decrease in the ice cover during this period.

In the Arctic, ice conditions reported by an ice observer were in good agreement with those found using Landsat data during the joint US/USSR sea ice experiment, BESEX (see Section 8.1), in February and March 1973 (Barnes *et al.*, 1974). Landsat was found to be very useful for monitoring sea ice conditions and tracking floe and lead patterns in an area south of St Lawrence Island in the Bering Sea under cloud-free conditions. Open water areas along the south shore of St Lawrence Island were observed to have caused stratus streaks to form revealing direct correlation of ice and atmospheric conditions. Visual observations on the BESEX flights corroborated interpretation of Landsat imagery.

Interannual variability of sea ice has always been of special interest to climatologists and others, but until the advent of satellite imagery it could not be reliably assessed. It was not until the launch of the ESMR on the Nimbus 5 satellite that the extent and changes of the global sea ice cover could be documented at frequent intervals regardless of weather or darkness. Analysis of the Nimbus 5 ESMR data of its first full year of operation (1973) showed that the

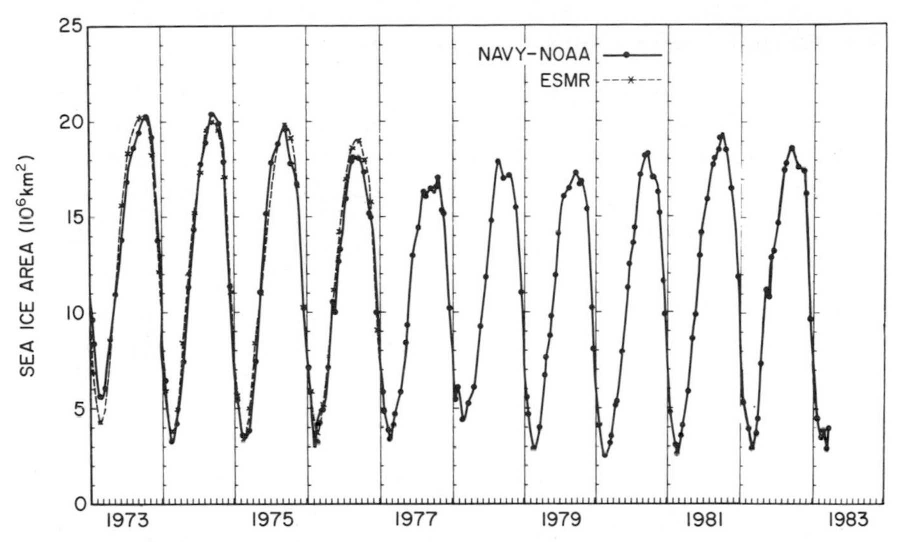

Fig. 8.5 Areal sea ice extent in the southern ocean as a function of time derived from US Navy/NOAA sea ice maps and Nimbus 5 ESMR observations (from Zwally *et al.*, 1983b.)

sea ice extent was quite different from what was previously believed both in the Arctic and Antarctic (Figs 7.13 and 7.14). Specifically, the sea ice extended 4° longitude farther east for some dates than had been predicted and open water areas that were previously thought to be ice covered were present north of Svalbard near Greenland. Early analysis of satellite-derived microwave maps also revealed large discrepancies in the distribution of multiyear ice in the north polar region (Gloersen *et al.*, 1975).

Recently, an atlas of the Antarctic sea ice has been published showing sea ice variability during 41 months from 1973 through 1976 (Zwally *et al.*, 1983a). This atlas presents monthly-averaged ESMR images from the Nimbus 5 satellite with color-coded T_Bs at intervals of 5 K, allowing monthly and yearly changes in sea ice cover and ice concentration to be studied. According to measurements made from the microwave maps, the extent of sea ice in the southern ocean typically varies from a minimum of 4×10^6 km^2 in February to a maximum of 20×10^6 km^2 in September. The ESMR data have also been used in conjunction with visible and infrared satellite data and ship reports as compiled in US Navy-NOAA maps in order to examine the interannual variations of the southern ocean over the 9-year period 1973–81 (Zwally *et al.*, 1983b and Zwally, 1984). The analysis showed a significant decrease in wintertime ice extent during the mid-1970s but a rebounding of the ice cover in subsequent years as seen in Fig. 8.5.

8.4 Sea ice concentration

The fraction of open water within sea ice is an important factor in the heat exchange between the atmosphere and the ocean. Within first-year sea ice, this fraction, F, can be calculated using single channel passive microwave data if the

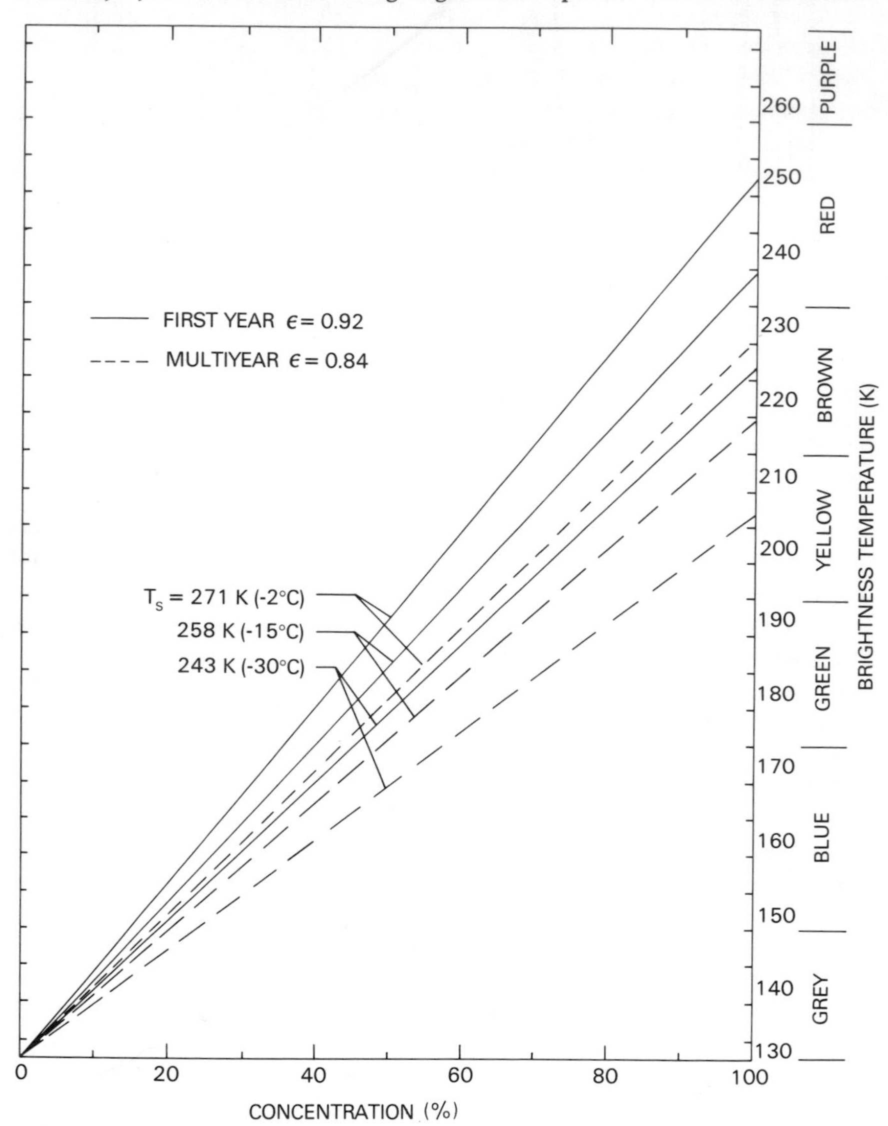

Fig. 8.6 Linear relationship between sea ice concentration and microwave brightness temperature for two values of emissivity and several values of the effective physical temperature of the ice radiating layer (from Zwally *et al.*, 1983a).

brightness temperature, T_B, and the ice surface temperature, T_s, are known. A simplified version of the algorithm used to determine F is given below:

$$F = (T_B - 0.92T_s)/(T_w - 0.92T_s) \quad (8.1)$$

where T_w is the T_B of smooth open water and 0.92 is the assumed emissivity of first-year sea ice at the 1.55 cm wavelength. Figure 8.6 shows the relationship between ice concentration and T_B in the southern ocean using Nimbus 5 ESMR data. Using single frequency data, ice concentrations are derived from linear interpolation between observed T_Bs of the open ocean and fully consolidated sea ice. Mean ice concentrations in the southern ocean range from 50% in the summer to 80% in the winter as determined from analysis of Nimbus 5 ESMR data (Zwally *et al.*, 1983a). Though the ESMR has been a very useful sensor in analyzing interannual as well as daily, weekly and monthly changes in sea ice concentration, improvement in the accuracy of ice concentration determination can be attained by employing two or more microwave channels.

The Nimbus 7 SMMR data have been used to improve the calculation of ice concentration by employing two microwave channels. Cavalieri *et al.* (1984) calculated multiyear ice fraction using a gradient ratio, GR, parameter:

$$\mathrm{GR} = (T_B(V, 0.8) - T_B(V, 1.7))/(T_B(V, 0.8) + T_B(V, 1.7)) \quad (8.2)$$

where the difference in the 0.81 cm and 1.7 cm vertically polarized T_Bs are divided by their sum. Use of the gradient ratio reduces the effects of physical temperature variations in the ice radiating layer. Thus an expression for calculating the multiyear ice fraction as given by Cavalieri *et al.* (1984) employs gradient ratio, a polarization parameter and ice physical temperature in the calculation of multiyear ice fraction.

MSS band 7 data have been used to calculate ice concentration values for comparison with ESMR data from the Nimbus 5 satellite (Comiso and Zwally, 1982). Figure 8.7 shows end-of-winter ice conditions as seen by Landsats 1 and 2 near the Mandheim coast of Antarctica (a), and near the Princess Astrid coast (b). The reflectivities of water and sea ice using MSS band 7 data can be distinguished using an interactive computer system (Fig. 8.8). These are assumed to be areas of first-year sea ice based on the microwave emissivity being approximately 0.92 as determined from analysis of Nimbus 5 ESMR data and appearance of the ice on the Landsat imagery. A comparison of ESMR-derived ice concentrations versus Landsat-derived ice concentrations is shown in Fig. 8.9. Acceptable agreement is obtained between the ESMR-derived ice concentrations and the MSS-derived ice concentrations (correlation coefficient = 0.744). However, the presence of narrow leads, small, unresolvable ice floes, thin ice and differential illumination cause considerable uncertainty in the ice concentrations that can be inferred from visible and near infrared sensors (Comiso and Zwally, 1982). This is especially true near the ice edge where many ice floes are smaller than the 80 m resolution of the Landsat MSS. Comiso and

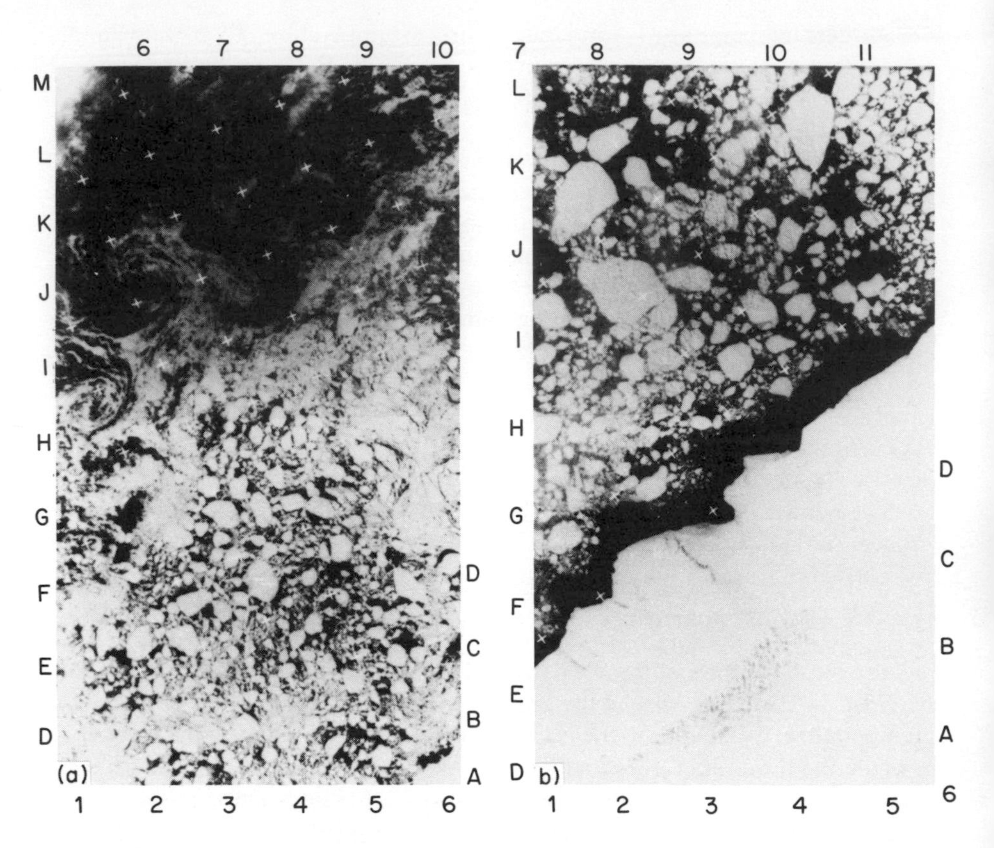

Fig. 8.7 MSS-7 near-infrared images from (a) Landsat 2, on 23 November, 1975 near the Mandheim coast of Antarctica, and (b) Landsat 1 on 13 November, 1973 near the Princess Astrid coast of Antarctica (from Comiso and Zwally, 1982).

Zwally (1982) used an integrating algorithm to extract ice concentration values from Landsat data to account for the unresolvable floes. This algorithm is similar to that used for extracting ice concentration values from brightness temperature data.

Areas of low (less than 10–15%) ice concentration are observable on both visible and microwave data. Such areas, known as polynyas, have been of considerable interest both historically and at present, and have been used by explorers to facilitate travel through ice-laden waters and to conduct seal and whale hunts. The presence of polynyas can have immense scientific interest as described in Section 8.1. Studies of the Bering Sea utilizing satellite imagery have shown that most winter ice formation occurs within several relatively small polynyas; these areas that have high rates of air–sea heat exchange and generate considerable amounts of new ice (Wadhams *et al.*, 1981). Ice formation in

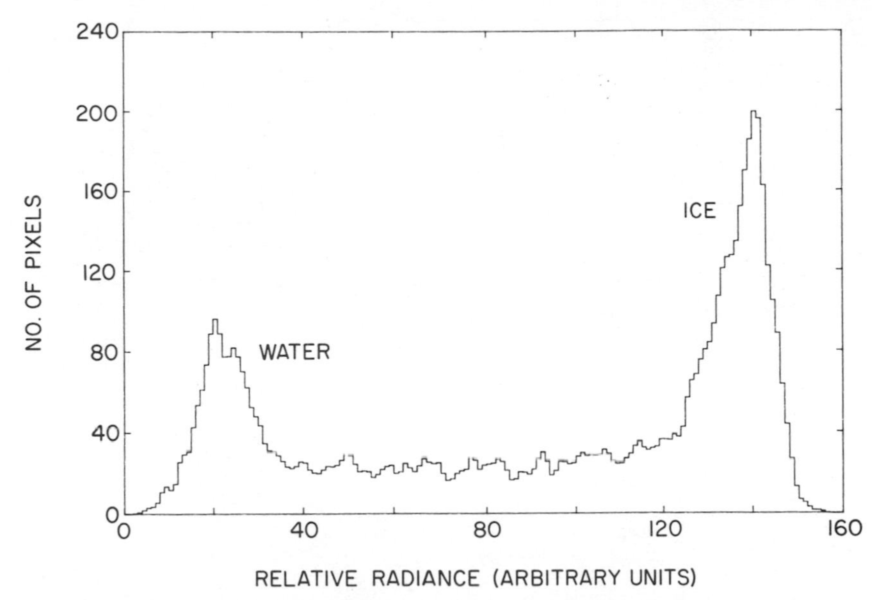

Fig. 8.8 Distribution of relative radiance for a typical Landsat area of the size of the ESMR pixel (adapted from Comiso and Zwally, 1982).

polynyas results in the formation of frazil ice (small disks of ice that form in the top few centimeters of the water surface). Ice core analyses in the Bering Sea show that 30–50% of the vertical extent of the ice cores analyzed consisted of frazil ice, suggesting that much of the ice had been formed in a polynya (Wadhams *et al.*, 1981).

The mechanics responsible for the formation of a polynya are often complex and unclear. It is likely that both oceanographic and atmospheric factors are influential in the formation and perpetuation of large polynyas both in the Arctic and Antarctic regions.

A large (200 km × 1000 km) open water area (less than 15% ice concentration) was observed on 1974 ESMR imagery of the southern ocean. Located at 0° longitude in the Weddell Sea, this large area of open water is known as the Weddell polynya. Though it was not seen on 1973 ESMR imagery (1973 was a year of unusually extensive ice), it was seen on satellite imagery during 1974, 1975 and 1976 but not between 1977 and 1983. Wind-enhanced divergence of ocean currents may contribute to upwelling of relatively warm water and result in low ice concentration in the area of the Weddell polynya (Zwally and Gloersen, 1977).

Carsey (1980) and Parkinson (1983) also used passive microwave data to describe the Weddell polynya, and Parkinson (1983) modeled various conditions which could lead to its formation. Both concluded that the formation

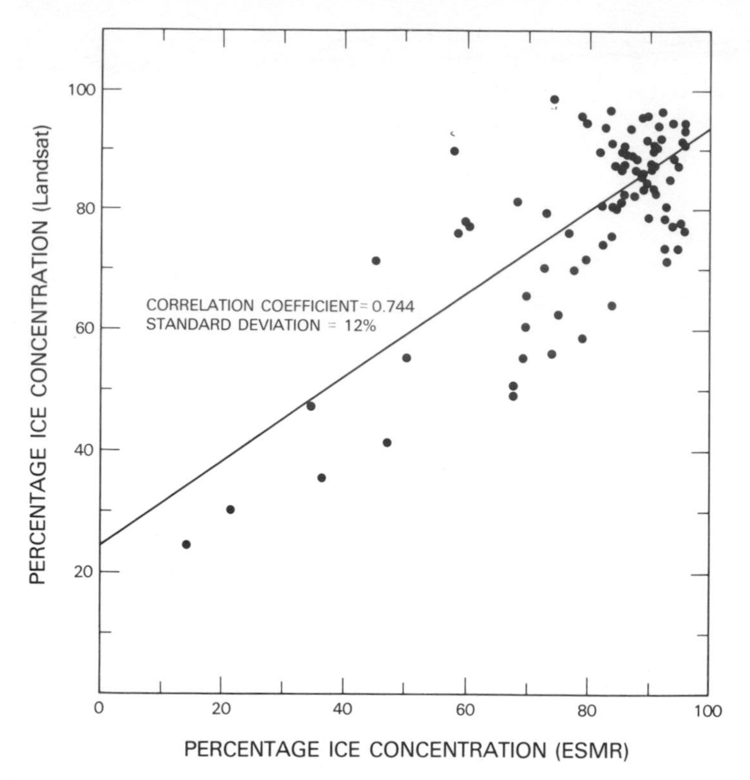

Fig. 8.9 Comparison of ice concentration derived from ESMR with those derived from Landsat (adapted from Comiso and Zwally, 1982).

and maintenance of the Weddell polynya is likely to be influenced by both ocean currents and winds. The Weddell polynya was observed to move in a westward direction over the period from 1974 to 1976 (Parkinson, 1983). Thus the polynya may be following the large scale oceanic circulation as observed by Martinson *et al*. (1981).

In the Arctic, Crawford and Parkinson (1981) used ESMR imagery to study the North Water, a polynya located in Smith Sound in northern Baffin Bay. Using 1973–76 observations, they employed the 193 K T_B isotherm on ESMR data to define the boundary of the North Water for their study. The wintertime North Water was found to open and close repeatedly, and often rapidly. The longest time span during which the polynya remained continuously open was 15 days in mid-February 1975 according to analysis of available Nimbus 5 ESMR data. In an effort to explain this rapid mid-winter opening and closing, Crawford and Parkinson (1981) noted that strong winds can open the polynya by forcing ice downwind whereas calm winds may allow the polynya to close because water freezes rapidly when in contact with the cold atmosphere.

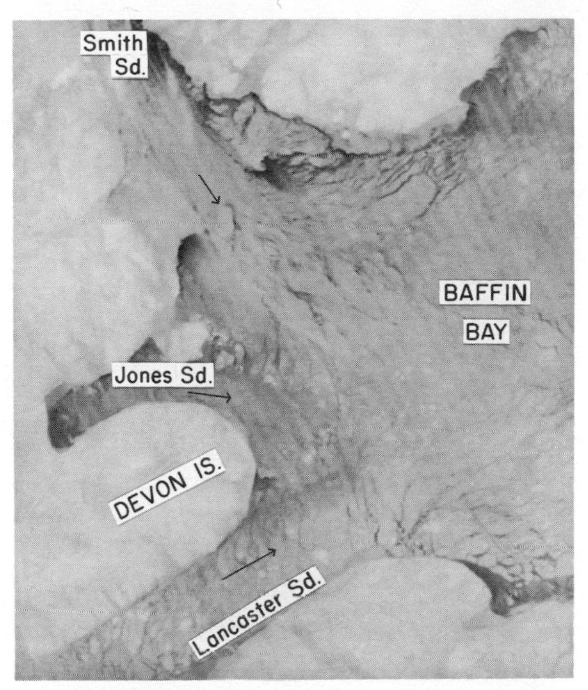

Fig. 8.10 NOAA (thermal infrared) image showing the direction of ice movement into Baffin Bay through Lancaster, Smith and Jones Sounds – 20 December, 1977 (courtesy of B. Dey, Howard University, Washington, D.C.).

However, as no strong correlations were found between winds and polynya opening events, it was concluded that ocean currents exert a strong influence as well.

The North Water polynya has also been studied using NOAA thermal infrared data for three winters (1974–75, 1975–76 and 1976–77) by Dey (1980). Gray levels, based on relative temperature differences, were used to establish the existence of ice cover during the polar night in these years. Strong temperature contrasts were observed between cold, thick sea ice and warmer open water and thin sea ice (Fig. 8.10). Open water and thin ice appears dark (relatively warm) on the images whereas thicker (colder) ice appears light in tone. Considerable heat flux from ocean to atmosphere and the accompanying higher evaporation in Smith Sound can be attributed to the North Water polynya. These factors encourage the higher frequency of winter cyclones in northern Baffin Bay (Dey, 1980).

Another persistent polynya has been observed using infrared satellite imagery in Terra Nova Bay, Antarctica. US Air Force Defense Meteorological Satellite Program (DMSP) images acquired daily with 3 km resolution were analyzed for the year 1979 (Kurtz and Bromwich, 1983). Even with cloudcover,

Table 8.3 Estimates of monthly mean extent of open water in the Terra Nova Bay polynya (from Kurtz and Bromwich, 1983)

Month 1979	*Mean area of open water (km²)*	*S*	S_x	*Number of* days	*Maximum area of open water* (km²)
March	3300	2700	1100	6	–
April	900	900	220	17	3000
May	1000	800	160	23	3000
June	1100	1000	230	19	4500
July	1100	600	120	25	3000
August	1400	1100	230	23	5000
September	1500	800	200	16	3400
October	1400	900	210	18	4500

Values used in calculating the means are based upon minimum estimates of open water extent on any given day.

numerous observations of the polynya were possible, as seen in Table 8.3. Open water appeared darker than sea ice on the imagery as on the NOAA imagery. Cold and thick sea ice has a low surface temperature and thus appears light gray on the imagery. Kurtz and Bromwich note that anomalously strong and stable katabatic (gravity) winds travelling down the Reeves glacier valley extend at least 25 km beyond the break in slope and are responsible for the failure of the sea ice in that area to consolidate. Water in the Terra Nova Bay will freeze in the absence of the katabatic winds. The average size of the polynya was 1000 km^2 during the polar night.

8.5 Sea ice movement

A satellite can provide a stable platform from which to observe and measure the movement of individual ice floes as well as larger-scale sea ice movements. Sea ice movement is controlled by ocean currents and winds. Landsat data have been employed to study sea ice movement during daylight hours in the Arctic and Antarctic. Radar data from aircraft and the Seasat satellite have proven extremely useful for measuring ice movement even during darkness and in inclement weather. NOAA VHRR and AVHRR data have been employed during the polar night.

Using Seasat SAR data, sea ice floes have been tracked and their drift velocities measured, both in a relative and absolute sense. Fu and Holt (1982) used Seasat SAR images to track the drift of an ice island which presumably broke off from the glacial ice of Ellesmere Island. Between 19 July and 7 October, 1978, the ice island (Fig. 8.11(a)) traveled 435 km as shown in Fig.

8.11(b) averaging 5.4 km day^{-1}. The greatest drift velocity (average 12.2 km day^{-1}) occurred between 28 September and 1 October. Figure 8.12 shows a computer-enhanced enlargement of the ice island shown in 8.11(a). It is a very bright object on the radar imagery because its surface roughness is sufficient to cause high backscatter and to saturate the dynamic response of the photographic film (Fu and Holt, 1982).

Leberl *et al.* (1979 and 1983) used satellite orbital data and ground control points to correlate with control points on SAR imagery for quantitative measurement of sea ice drift. Ground co-ordinates were generated by using: (1) image co-ordinates of the same point on all images, (2) ground control points of a known point (e.g. a station on an ice floe), (3) tie points between overlapping images and (4) sensor orientation and position parameters (Leberl, 1983). In the Beaufort Sea, Leberl *et al.*, (1983) measured sea ice movement from 7 Seasat overpasses. The average velocity was 6.4 ± 0.5 km day^{-1}. This velocity is roughly comparable with that obtained by Fu and Holt (1982) for the ice island. Elachi (1980) calculated an average ice velocity of up to 15 km day^{-1} using two radar images of the Beaufort Sea just north of Banks Island in Canada. The images were acquired 3 days apart. As an ice feature changes with time, it becomes increasingly difficult to track, thus limiting the accuracy of sea ice movement studies.

Simultaneous SLR and visual observations of sea ice conditions were made by Dunbar (1975) in Nares Strait, off the northwest coast of Greenland. Flights were conducted in January, March and August of 1973 using a real aperture X-band radar. Multiyear sea ice (as determined by in-flight ice observers) was found to be especially variable in appearance on the X-band imagery. In addition to the Nares Strait study area, a March 1973 flight was conducted from Alert to the North Pole. The characteristics of the ice, as observed on radar imagery, were quite different from those in Nares Strait. SLAR data were used to track sea ice floes in Nares Strait using an aircraft platform. During a 3-day period in October of 1976, one ice floe was observed to travel at a top speed of 0.81 m sec^{-1} (Dunbar, 1979).

Using Landsat MSS data, winds have been found to influence sea ice transport. Dey (1981) used MSS data to track floes or collections of floes using sequential imagery in Smith Sound in northern Baffin Bay, Canada. Ice speeds were determined. The winds in the area are influential in sea ice transport. Ito and Muller (1982) measured southbound sea ice movement through Smith Sound and determined a mean speed of 4.3 km d^{-1} with the greatest speed measured at 34.0 km d^{-1}. They found that ice movement as derived from Landsat data could be used to evaluate the wind-field of the area.

Because of overlapping orbital swaths in the polar regions, Landsats 1, 2 and 3 data could be acquired (cloudcover permitting) for periods of more than 5 days in a row, thus permitting the tracking of individual ice floes. Rango *et al.* (1973) calculated the average velocity of ice floe movement over a 5-day period in

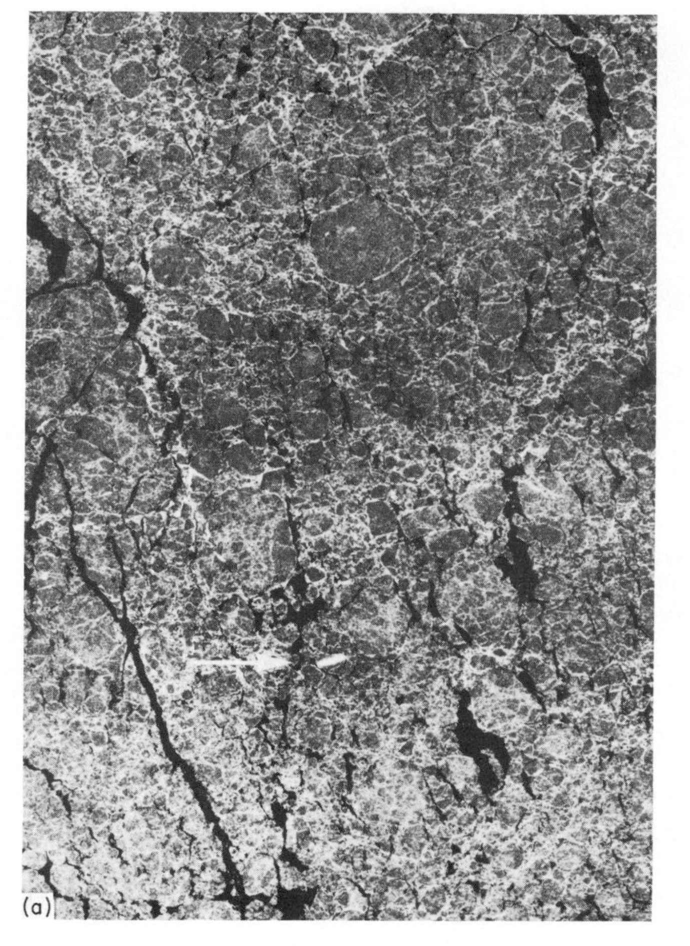

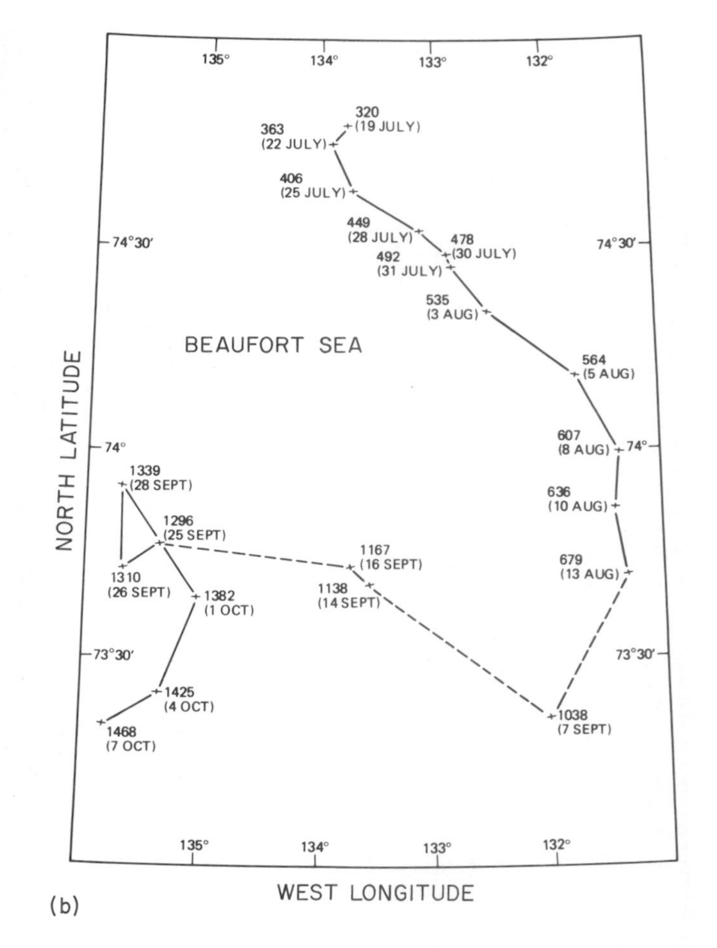

Fig. 8.11 Seasat SAR image of pack ice in the central Beaufort Sea (a). Note ice island (bright feature) at arrow in (a), and drift motion of the ice island in (b). The crosses indicate the position and orientation of the ice island plus the Seasat revolution number and date. The dashed lines indicate a significant gap in the SAR coverage of this feature (adapted from Fu and Holt, 1982, courtesy of Jet Propulsion Laboratory).

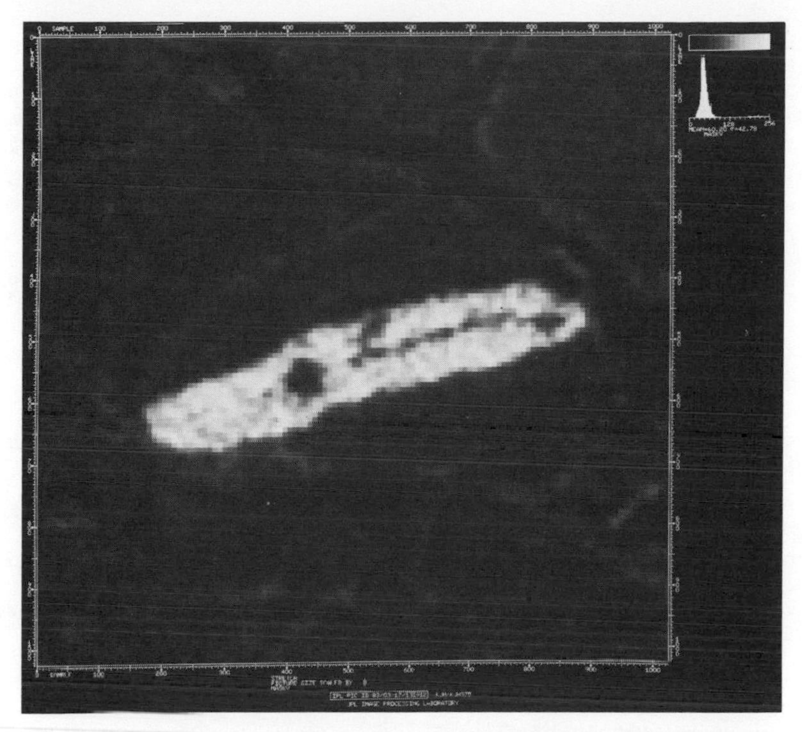

Fig. 8.12 Computer-enhanced enlargement of the ice island shown in Fig. 8.11(a) (courtesy of Jet Propulsion Laboratory).

August of 1972 to be 8.5 km d^{-1} in Hendriksen Strait in the Queen Elizabeth Islands of Arctic Canada.

Near-shore pack ice movement has been successfully measured with Landsat data to reveal information on sea ice deformation and drift. Landsat near-infrared data were analyzed and used to confirm a theory of ice motion in the shear zone north of Alaska (Hibler *et al.*, 1974). The shear zone is a near-shore zone of ice which is subjected to divergent forces. Using Landsat data of a 4-day period in March 1973, strain calculations were made. The pack ice was found to behave as a coherent mass rotating clockwise with slippage in a narrow (50 km wide) region.

In this chapter we have attempted to show that remote sensing is highly suited to the study of sea ice. Analysis of the role of sea ice in oceanic processes can be accomplished using remote sensing techniques. Additionally, analysis of sea ice movement, concentration, type, variability and age are all important for routing ships through ice-laden waters.

References

Barnes, J.C., Bowley, C.J., Chang, D.T. and Willard, J.H. (1974) Application of satellite visible and infrared data to mapping sea ice. In *Advanced Concepts in the Study of Snow and Ice Resources*, (eds H.S. Santeford and J.L. Smith) US International Hydrological Decade, National Academy of Sciences, Washington, DC, pp. 467–76.

Bryan, M.L. (1976) Interpretation key for SAR (L-band) imagery of sea ice, *Proceedings of the American Society of Photogrammetry Fall Convention, 28 September – 1 October 1976, Seattle, WA*, pp. 406–35.

Campbell, W.J., Weeks, W.F. Ramseier, R.O. and Gloersen, P. (1975) Geophysical studies of floating ice by remote sensing. *J. Glaciol*, **15**, 305–28.

Campbell, W.J., Wayenberg, J. and Ramseyer, J.B. *et al.* (1978) Microwave remote sensing of sea ice in the AIDJEX main experiment. *Boundary-Layer Meteorol.*, **13**, 309–37.

Carsey, F.D. (1980) Microwave observations of the Weddell Polynya. *Mon. Weather Rev.*, **108**, 2032–44.

Cavalieri, D.J., Gloersen, P. and Campbell, W.J. (1984) Determination of sea ice parameters with the Nimbus-7 SMMR. *J. Geophys. Res.*, **89**, 5355–69.

Comiso, J.C. and Zwally, H.J. (1982) Antarctic sea ice concentrations inferred from Nimbus-5 ESMR and Landsat imagery. *J. Geophys. Res.*, **87**, 5836–44.

Crawford, J.P. and Parkinson, C.L. (1981) Wintertime microwave observations of the North Polar polynya. In *Oceanography from Space* (ed. J.F.R. Gower) Plenum Publishing Corporation, New York, pp. 839–44.

Dey, B. (1980) Applications of satellite thermal infrared images for monitoring North Water during the periods of polar darkness. *J. Glaciol.* , **25**, 425–38.

Dey, B. (1981) Monitoring winter sea ice dynamics in the Canadian Arctic with NOAA TIR images. *J. Geophys. Res.*, **86**, 3223–35.

Dunbar, M. (1975) Interpretation of SLAR imagery of sea ice in Nares Strait and the Arctic Ocean. *J. Glaciol.*, **15**, 193–213.

Dunbar, M. (1979) Fall ice drift in Nares Strait, as observed by sideways-looking airborne radar. *Arctic*, **32**, 283–307.

Elachi, C. (1980) Spaceborne imaging radar: geologic and oceanographic applications. *Science*, **209**, 1073–82.

Fu, L.L. and Holt, B. (1982) *Seasat views oceans and sea ice with synthetic aperture radar*, Jet Propulsion Laboratory, Pasadena, CA, JPL publication 81–120.

Gloersen, P., Wilheit, T.T., Chang, T.C. and Nordberg, W. (1975) Microwave maps of the polar ice of the Earth. In *Climate of the Arctic*, (eds G. Weller and S.A. Bowling), Proceedings of the Twenty-Fourth Alaska Science Conference, 15–17 August, 1973, Geophysical Institute, University of Alaska, Fairbanks, AK, pp. 407–14.

Gow, A., Ackley, S.F. Weeks, W.F. and Gavoni, J.W. (1982) Physical and structural characteristics of Antarctic sea ice. *Ann. Glaciol*, **3**, 113–17.

Hengeveld, H.G. (1974) Remote sensing applications in Canadian ice reconnaissance. In *Advanced Concepts in the Study of Snow and Ice Resources*, (eds H.S. Santeford and J.L. Smith), US International Hydrological Decade, National Academy of Sciences, Washington, DC, pp. 504–12.

Hibler, W.D. III, Ackley, S.F. and Crowder, W.K. *et al.* (1974) Analysis of shear zone deformation in the Beaufort Sea using satellite imagery. In *The Coast and the Shelf of the Beaufort Sea* (eds J.C. Reed and J.E. Sater) Proceedings of a symposium on Beaufort Sea Coast and Shelf Research, Arctic Institute of North America, Arlington, VA, pp. 285–96.

Ito, H. and Muller, F. (1982) Ice movement through Smith Sound in northern Baffin Bay, Canada observed in satellite imagery. *J. Glaciol.*, **28**, 129–43.

Ketchum, R.D. (1983) Dual frequency radar ice and snow signatures. *J. Glaciol.*, **29**, 286–95.

Kurtz, D.D. and Bromwich, D.H. (1983) Satellite observed behavior of the Terra Nova Bay polynya. *J. Geophys. Res.*, **88**. 9717–22.

Langham, E.J. (1982) RADARSAT – Canada's program for operational remote sensing. *Can. J. Remote Sensing*, **8**, 29–37.

Leberl, F. (1983) Photogrammetric aspects of remote sensing with imaging radar. *Remote Sensing Rev.* **1**, 71–158.

Leberl, F., Bryan, M.L. and Elachi, C. *et al.* (1979) Mapping of sea ice and measurement of its drift using aircraft Synthetic Aperture Radar images. *J. Geophys. Res.*, **84**, 1827–35.

Leberl, F., Raggam, J., Elachi, C. and Campbell, W.J. (1983) Sea ice motion measurements from SEASAT SAR images. *J. Geophys. Res.*, **88**, 1915–28.

Lyden, J.D., Burns, B.A. and Maffett, A.L. (1984) Characterization of sea ice types using synthetic aperture radar. *IEEE Trans. Geosci. Remote Sensing*, **GE-22**, 431–9.

Martinson, D.G., Killworth, P.D. and Gordon, A.L. (1981) A convective model for the Weddell polynya. *J. Phys. Oceanogr.*, **11**, 466–88.

Parkinson, C.L. (1983) On the development and cause of the Weddell Polynya in a sea ice simulation. *J. Phys. Oceanogr.*, **13**, 501–11.

Pounder, E.R. (1965) *Physics of Ice*, Pergamon Press, Oxford, England.

Rango, A., Greaves, J.R. and DeRyke, R.J. (1973) Observations of Arctic sea ice dynamics using the Earth Resources Technology Satellite (ERTS-1). *Arctic*, **26**, 337–9.

Wadhams, P., Martin, S., Johannessen, O.M. *et al.* (1981) *MIZEX, A Program for Mesoscale Air – Ice – Ocean Interaction Experiments in Arctic Marginal Ice Zones, 1. Research Strategy*, US Army Cold Regions Research and Engineering Laboratory, Hanover, NH, CRREL SP 81-19.

Weeks, W.F. (1981) Sea ice: the potential of remote sensing. *Oceanus*, 24, 39–48.

Zwally, H.J. and Gloersen, P. (1977) Passive microwave images of the polar regions and research applications. *Polar Rec.*, **18**, 431–50.

Zwally, H.J., Comiso, J.C. and Parkinson, C.L. *et al.* (1983a) *Antarctic Sea Ice, 1973–1976: Satellite Passive Microwave Observations*, National Aeronautics and Space Administration, Washington, DC, NASA SP-459.

Zwally, H.J., Parkinson, C.L. and Comiso, J.C. (1983b) Variability of Antarctic sea ice and changes in carbon dioxide. *Science*, **220**, 1005–12.

Zwally, H.J. (1984) Observing polar ice variability. *Ann. Glaciol.*, **5**, 191–98.

Index

Page numbers which refer to figures have been italicized.